Fundamentals of Elementary Algebra

McGraw-Hill Basic Skills System

Tools for Learning Success

Tests

STUDY SKILLS	READING	VOCABULARY	SPELLING	WRITING	MATH
Study Skills	Efficient Rates	Vocabulary Skills	Spelling Skills	Writing Skills 1	Arithmetic
Problem Solving	The Main Idea		Basic Spelling	Writing Skills 2	Elementary Algebra
Listening & Notetaking	Significant Facts			Paragraph Patterns	Intermediate Algebra
Underlining	Organization				
Library & Reference Skills	Main Idea in Science				
	Critical Reading				
	Study-Type Reading				
	Skimming and Scanning				

Materials

Editor's Introduction

This book is one of three parts in a series devoted to instruction in mathematics skills. This mathematics series, in turn, is part of a larger system of instructional materials—the McGraw-Hill Basic Skills System: Tools for Learning Success. Designed at the University of Minnesota Reading and Study Skills Center, the Basic Skills System is aimed at college-bound high school students, and junior college and college students who need to improve those skills necessary for academic success. The system consists of *tests* to determine instructional needs and *materials* designed to meet those needs, plus instructor's manuals to explain the tests and materials and the relationship between them. The purpose of the *tests* is to find out what instruction a student needs in basic skills, and the purpose of the *materials* is to give him that instruction. Each student gets what he needs without wasting time on unnecessary tasks.

Six basic skill topics—study skills, reading, vocabulary, spelling, writing, and mathematics—are covered, and two tests (A and B forms) are provided for each topic. Subscales on the tests are matched to accompanying instructional materials: thus a student with a low score on one or more subscales gets instruction in the corresponding skill. The second form of the test may be used to evaluate progress after instruction.

The materials in the Basic Skills System have been field-tested and revised to provide the best possible results. While most of the materials are self-instructional programs, the tests, of course, are designed for supervised administration. These testing instruments have been carefully developed and standardized by California Test Bureau. The latest research techniques and procedures have been utilized to ensure the highest possible validity and reliability.

The instructional materials are designed to be used separately, if desired, and can be purchased as single units. Most of the materials are suitable for adoption as textbooks in such basic skill courses as Freshman English, Communications, How to Study, Vocabulary Development, and Remedial or Developmental Mathematics. Individualized diagnosis and instruction are optional in such settings.

Alton L. Raygor
Consulting Editor
University of Minnesota

Fundamentals of Elementary Algebra

A Program for Self-instruction

MICHAEL ERAUT
Centre for Educational Technology
University of Sussex, England

McGRAW-HILL BOOK COMPANY
New York St. Louis San Francisco Düsseldorf London
Mexico Panama Sydney Toronto

Fundamentals of Elementary Algebra
A Program for Self-instruction

07-051402-X

2 3 4 5 6 7 8 9 0 V B V B 7 9 8 7 6 5 4 3 2 1

McGraw-Hill Basic Skills System:
Design Director: Barbara Bert
Production Supervisor: Adam Jacobs
Editing Supervisor: Antonia Stires
Editor: James Wallace

The author developed this book in connection with an urban education project involving approximately 200 marginally prepared and specially enrolled freshman students at the University of Illinois at Chicago Circle while he was on the staff of the Office of Instructional Resources.

Preface

I began this book because I believed that the problem of freshman mathematics could be considerably simplified by specifically catering for variations in student achievement rather than trying to compromise with them. It is virtually impossible to adapt traditional courses to such a wide range of student needs; so I set out to design a self-instructional course which students could start at any point. The next problem was how to help the students decide which parts of the course they needed, and this led me to the rather complex system of diagnostic tests described on page ix.

The first version of each chapter was tried out with a very small group of students and carefully discussed with them so that I could use their experience and advice to produce an improved version. Sometimes I had to do this two or three times before I was satisfied. Then the testing with larger numbers of students began, and this led to two further revisions. I also had four mathematics professors independently examine the material in detail, and I have incorporated most of their suggestions. Nearly all the testing was carried out at the University of Illinois at Chicago with students scoring in the lowest 20% on the Mathematics Placement Test. Students placed higher would still derive considerable benefit, but the aim of the testing was to make sure that the book was good enough to teach even the weakest students. The results obtained with the last version prior to publication show good evidence of its effectiveness with these students; and it has been improved since then. These results were obtained with nearly 80 freshmen at the University of Illinois:

Average score on diagnostic tests at beginnings of chapters	28%
Average score on final tests at ends of chapters	89%

The time taken varies greatly from one student to another. In general the shorter sections take 10 to 40 minutes and the longer sections 30 to 90 minutes to complete; but some students take most of the sections and some take very few.

Michael Eraut

To the Student

THE PURPOSE OF THIS BOOK

This book was written because not all students are the same. You differ in the way you learn, in your knowledge of mathematics, and in your ability to apply your knowledge to problems in science, social science, and technology. These differences make it particularly difficult to cater for your needs when you first arrive in college. There may be three or four mathematics courses you can take, but a course represents a lot of work. You won't want to take Math 100 when you only need the last 2 weeks of it; yet the last 2 weeks of Math 100 may be essential background for Math 101. If Math is one of your weaker subjects, you may even find that Math 100 starts at a level too high for you and assumes that you know things which you can only hazily remember from high school and never properly understood. If your first college course is too easy, you get bored; if it is too difficult, you get frustrated. The purpose of this book is to solve that problem.

The book has two main components, diagnostic tests and the self-instructional program. The purpose of the tests is to make sure that (1) you understand the fundamentals and (2) you skip all the sections of the book you don't need. Then you won't waste any time. The program itself will provide you with learning sequences in the form of numbered questions and answers called "frames." These have been thoroughly tested and revised so that you can learn to a very high degree of proficiency all the material in the book that you don't already know.

Since you will probably be meeting much of the material for the second time, this book attempts to approach most of it from a fresh viewpoint and to make it meaningful to students who have taken either a modern or a traditional course in high school. At the same time it tells you exactly where you are going and why. It leads you up to rules through sequences of examples, often allowing you to deduce them for yourself first, and it makes sure that you are convinced of their validity. But it does not attempt many proofs. The overall purpose is to find out what you need, to give it to you as quickly as possible, and to encourage an understanding of the fundamental principles in addition to developing a proficiency in applying them.

HOW TO USE THE TESTS

Each chapter in this book has a diagnostic test at the beginning and a final test at the end. The chapters are divided into sections, each with a pretest at the beginning and a posttest at the end. You mark all the tests yourself, using the answers provided. This may seem to you like a lot of tests even though some of them are very short, but the point of the tests is to help you find your best way through the book. It is impossible to adapt to your individual needs without them. You will understand this better when you are halfway through the book and have a better grasp of how the system works. Meanwhile let us follow Edward, a typical student, in his progress through a chapter. Edward has some previous knowledge but is not fully proficient. The diagram below shows his path through the chapter and is followed by an explanation of how the testing system allocated it.

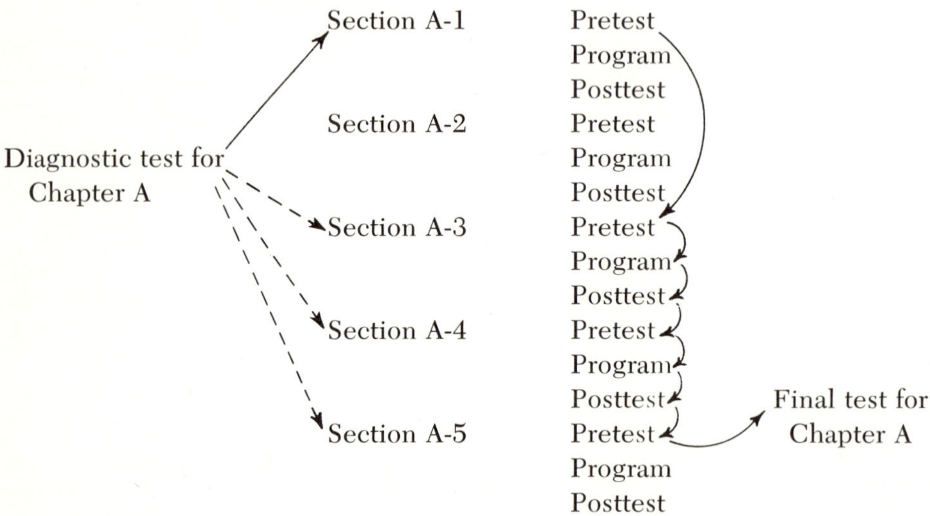

Edward began by taking the diagnostic test and marked it himself as follows (✓ indicates an error)

1.	✓	4.		6.	✓	9.	✓
2.		5.		7.		10.	✓
3.				8.	✓		

He then followed the instructions below and circled Secs. A-1, A-3, A-4, and A-5:

	section	page
If you made any mistakes in Questions 1 to 3, circle Sec. A-1.	(A-1)	000
If you made any mistakes in Question 4 or 5, circle Sec. A-2.	A-2	000
If you made any mistakes in Questions 6 to 8, circle Sec. A-3 *and* Sec. A-4.	(A-3) (A-4)	000 000
If you made any mistakes in Question 9 or 10, circle Sec. A-5.	(A-5)	000

This means that the diagnostic test has assigned Edward to Secs. A-1, A-3, A-4, and A-5 (but not A-2).

Edward now takes the pretest for Sec. A-1 and gets it all correct (his wrong answer to Question 1 of the diagnostic test was a careless mistake which he didn't repeat). This enables him to skip Sec. A-1.

Since the next section assigned is A-3, Edward takes Pretest A-3. He gets two out of six, and so he stays with A-3, works through the program, and takes the posttest. He gets six out of six on the posttest (the typical student *does* learn nearly all the material!).

Then Edward takes the pretest for Sec. A-4 and gets none of them right. He works through the program for A-4 and gets five out of six on the posttest.

When he takes the pretest for Sec. A-5, he answers all the questions correctly, and so he skips Sec. A-5. He couldn't have passed this test at the beginning, but now that he has worked through two sections he has remembered some of his high school work (and probably understands it better).

Edward is now ready to take the final test at the end of the chapter. He gets nine out of ten because he made a careless mistake.

The tests also serve two other purposes. They help you to see in advance just what you are going to learn, and they help you to check up afterward on your success. A diagnostic test tells you what a chapter contains, and a pretest tells you what a section contains. Then the posttest and the final test help you to check that you have learned everything. You should make very few mistakes on these as the book has been well tested and thoroughly revised.

Do not worry if you do not remember everything in this introductory section. You will receive detailed instructions as you work through the book. The object of this section is to give an overview. But make sure that you follow the instructions exactly. If they say that you need to go through a program, they are right. This has been carefully tested because I did not want to give students any unnecessary work. But during the testing of this book, those students who failed to follow the instructions nearly always did worse than the others.

HOW TO USE THE PROGRAMS

Each section of this book is a self-instructional program, which has been carefully designed to lead you through the material efficiently and effectively. The content is broken down into a series of steps, or frames, and each frame contains one or more questions and the appropriate answers. The material is arranged like Frames 1 to 4 on the next page.

1 Which section pretest was skipped by Edward in the diagram on page x?

Sec. A-2

2 Why was he allowed to skip this pretest?

Because he got both Questions 4 and 5 of the diagnostic test correct.

3 Edward was also able to skip the frames for Secs. ______ and ______ because he passed the section pretests.

A-1 A-5

4 He worked through the frames for Secs. ______ and ______.

A-3 A-4

Nearly all students find it useful to cover the answers to each frame while they are working on the questions. An easy way to do this is to use the mask card provided with this book and to position it with its top against the line in the middle of the frame. After you have studied the frame, write out your answer in full on a separate piece of paper. Then move the mask down the page to the next line, exposing the answer, and compare your own answer with the one given in the book.

You are unlikely to be wrong more than about 10% of the time, but when you are, make sure that you can understand how the answer in the book was derived. If you don't do this, you will find that later material begins to get very difficult because you are not sure of the fundamentals. In any case your performance on the posttest at the end of the section will tell you if you have any serious misunderstandings and need to review.

Michael Eraut

Contents

Contents

Chapter 1
The Language of Algebra

The new terminology you will learn in this chapter provides a sound basis for the rest of the book. In Secs. 1-1 and 1-2 you will meet new signs, $\neq$, $<$, $>$, $\leqslant$, $\geqslant$, $\sqrt{\ }$, and some new kinds of number—irrational numbers and real numbers. Section 1-3 introduces the concept of a set and gives you practice in writing sets in two kinds of notation. This leads in Sec. 1-4 to the concepts of subset, union, and intersection. In Sec. 1-5 you will learn the meaning of the terms variable, domain, expression, open sentence, equality, inequality, equation, and inequation. Section 1-6 explains the concept of a solution of an open sentence and gives you practice in finding the solution sets of simple equations and inequations in one variable. It also shows you how to graph them.

This chapter requires the ability to add, subtract, multiply, and divide simple integers and to graph rational numbers.

DIAGNOSTIC TEST 1

1. Which of the following statements are true?

   a. $-18 \neq -9 \times 2$ *b.* $-3 < -7$
   c. $-4 \geqslant -16 \div 4$ *d.* $-5 \leqslant -3 \times (-2)$

2. If k is an integer and $-3 \leqslant k < 2$, what values can k have?

3. Which of the following numbers

   $\sqrt{30} \qquad \frac{7}{-3} \qquad -\sqrt{16} \qquad \frac{0}{5} \qquad \frac{-6}{3} \qquad -4$

   are:

   a. real numbers *b.* rational numbers
   c. irrational numbers *d.* integers

4. Write the following sets in element notation:

   a. $\{n \mid n \text{ is an integer and } n < 4\}$
   b. $\{-2, -1, 0, 1, 2\} \cup \{1, 2, 3, 4\}$
   c. $\{3, 4, 5, 6\} \cap \{5, 7, 8, 9\}$

5. Write the following sets in property notation:

   a. $\{-17, -16, -15, -14, -13\}$
   b. $\{n \mid n$ is a positive rational number$\} \cap \{m \mid m$ is a negative rational number$\}$
   c. $\{y \mid y$ is a real number$\} \cup \{z \mid z$ is an integer$\}$
   d. $\{p \mid p$ is a negative integer$\} \cap \{q \mid q$ is a rational number$\}$

6. Which of the following sentences are closed, and which are open?

   a. $x + 3 = 7$
   b. $w - 1 = w - 5$
   c. $(2 \times 4) + 3 = 10$
   d. $12 > y + 5$
   e. $m + 1 < m + 2$
   f. $-1 \times (-5) = 5$

7. Which of the sentences in Question 6 are true?

Suppose x is a variable over the set of numbers $\{1, 3, 5, 7, 9, 11\}$:

8. What is the domain of x?

9. Graph the solution set of the open sentence $x + 7 = 14$.

10. Graph the solution set of the open sentence $x > 5$.

Give the solution sets of the following equations (*a*) when the domain of x is $\{0, 2, 4, 6, 8, 10\}$ and (*b*) when the domain of x is $\{x \mid x$ is a real number$\}$:

11. $x + 7 = 10$ *a.* ________   *b.* ________

12. $x + 7 = x + 8$ *a.* ________   *b.* ________

13. $x + 8 = x + 8$ *a.* ________   *b.* ________

When you have finished the diagnostic test, correct it yourself using the answers on page 56. Then check all the questions you got wrong on the list below:

1. ______	3*a*. ______	4*a*. ______	6. ______	9. ______
2. ______	3*b*. ______	4*b*. ______	7. ______	10. ______
	3*c*. ______	4*c*. ______	8. ______	11*a*. ______
	3*d*. ______	5*a*. ______		11*b*. ______
		5*b*. ______		12*a*. ______
		5*c*. ______		12*b*. ______
		5*d*. ______		13*a*. ______
				13*b*. ______

If you made no mistakes at all, you need not do any work in this chapter and should go straight on to Chap. 2, First-degree Equations in One Variable, which begins on page 61. But if you made any mistakes (even if it was only *one* mistake), follow these instructions:

	section	page
If you made any mistakes in Question 1 or 2, circle Sec. 1-1.	1-1	4
If you made any mistakes in Question 3, circle Sec. 1-2.	1-2	14
If you made any mistakes in Question 4 or 5, circle Secs. 1-3 and 1-4.	1-3 1-4	21 29
If you made any mistakes in Questions 6 to 8, circle Sec. 1-5.	1-5	38
If you made any mistakes in Questions 9 to 13, circle Sec. 1-6.	1-6	47

You should now work through the sections whose numbers you have circled, taking them in numerical order. The sections begin on the pages indicated.

Section 1-1
Inequalities

Do this section if you made a mistake in Question 1 or 2 of Diagnostic Test 1. Begin by taking the pretest below, which will show you how much of Sec. 1-1 you already know and what you will be able to do by the time you have finished it.

PRETEST 1-1

Which of the following statements are true?

1. $-12 \neq (-4) \times 3$
2. $-10 < -12$
3. $12 > 7 + 9$
4. $8 \geqslant 4 \times 2$
5. $6 \neq 7 + 1$
6. $2 \leqslant -8 \div (-4)$
7. If $y = 4$, is $y \leqslant 4$? ______ (yes/no)
8. If $y > -3$, is $y \geqslant -3$? ______ (yes/no)
9. If k is an integer and $-5 < k \leqslant 0$, what values can k have?
10. Express the following as a single statement using $<$ or $\leqslant$ signs: n is less than -2 and greater than or equal to -11.

When you have finished the pretest, correct it yourself using the answers on page 57. Then check all the questions you got wrong on the list below:

1.	______	6.	______
2.	______	7.	______
3.	______	8.	______
4.	______	9.	______
5.	______	10.	______

If you made any mistakes (even if it was only *one* mistake), read the introduction and go through the program.

If you made no mistakes at all, skip Sec. 1-1, because you already know its material, and go on to the next section whose number you circled on page 3. If no further numbers were circled, take Final Test 1 (page 54).

Introduction

This section introduces you to six signs for expressing the relationship between numbers, $=$, $<$, $>$, $\neq$, $\leqslant$, $\geqslant$, and shows you how to represent a range of numbers by statements of the form $3 < k < 8$ or $5 \leqslant n \leqslant 9$.

FRAMES FOR SECTION 1-1

1 The sign $=$, meaning "equal to" will be familiar to you from arithmetic, but its opposite, $\neq$, meaning "not equal to," may be less familiar. Which of the following statements are true?

a. $2 \times 2 = 4$
b. $3 \times 2 \neq 5$
c. $4 \times 3 \neq 12$
d. $6 \times 2 = 14$

---

a and *b*

---

2 Which of the following statements are true?

a. $5 \times 2 \neq 10$
b. $6 \times 3 = 18$
c. $11 \times 5 = 45$
d. $7 \times 4 \neq 32$

---

b and *d*

---

3 Which of the following statements are true?

a. $4 \times 0 \neq 0$
b. $5 \times (-3) = 15$
c. $-4 \times (-3) \neq -12$
d. $-2 \times 6 = -12$

---

c and *d*

---

4 Two other signs used in algebra are $<$, meaning "less than," and $>$, meaning "greater than." Which of the following statements are true?

a. $3 \times 4 > 13$
b. $5 \times 3 < 13$
c. $6 \times 3 > 16$
d. $2 \times 4 < 9$

c and *d*

5 Notice that the narrow, or smaller, end of the symbol always points toward the numeral for the (larger/smaller) __________ number. (Look at the answer to Frame 4 if you need to.)

smaller

6 Which of the following statements are true?

a. $3 \times 8 > 21$
b. $2 \times 6 > 13$
c. $7 \times 9 < 65$
d. $3 \times 12 > 40$

a and *c*

7 In comparing two numbers, the greater number is always to the (right/left) __________ of the smaller number on the number line.

right

8 Which of the following statements are true?

a. $7 < -1$
b. $-\frac{1}{4} > -\frac{1}{2}$
c. $-3 > -2$
d. $-6 < -5$

b and *d*

9 Insert < or > to make each of the following statements true:

a. $-\frac{3}{4} \quad -\frac{1}{2}$
b. $-1 \quad 0$
c. $-2\frac{1}{2} \quad -2\frac{1}{4}$
d. $-3 \quad -7$

a. < *b.* < *c.* < *d.* >

10 We now have four symbols which express a relationship between two numbers: $=$, $\neq$, $>$, $<$. The meanings of these symbols can be pictured on the number line.

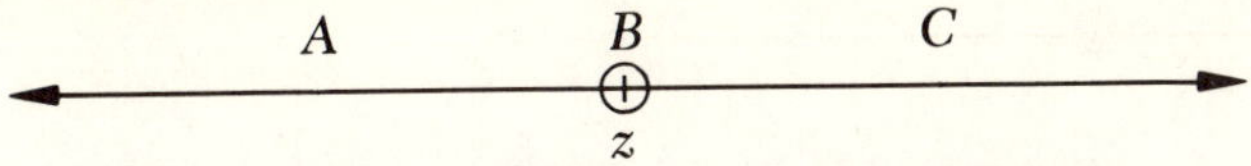

a. If $y = z$, then y is in the region ($A/B/C$) ______.
b. If $y < z$, then y is in the region ($A/B/C$) ______.
c. If $y > z$, then y is in the region ($A/B/C$) ______.
d. If $y \neq z$, then y is either in region ______ or in region ______ but not in region ______.

---

a. B *b.* A c. C *d.* A; C; B

---

11 If we shade the region of the line in which y belongs, we can consider these four cases.

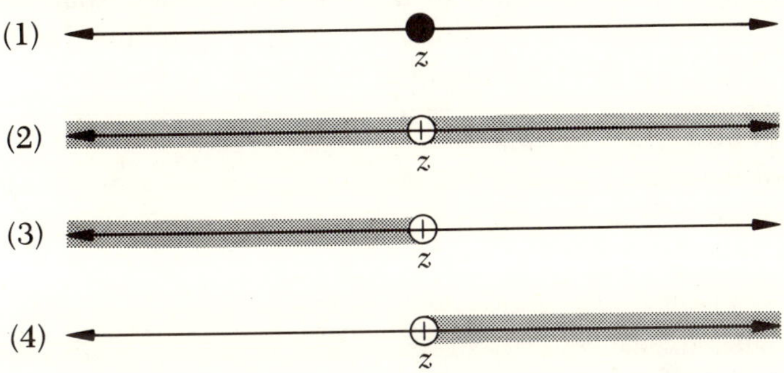

Case 1 corresponds to the situation where $y = z$.
a. Case 2 corresponds to the situation where y ______ z.
b. Case 3 corresponds to the situation where y ______ z.
c. Case 4 corresponds to the situation where y ______ z.

---

a. $\neq$ *b.* $<$ c. $>$

---

12 You will notice that in the picture of case 2, $y \neq z$, two of the three regions of the line are shaded. We shall now look at two other cases which have two areas shaded.

(5) 

(6) 

In case 5 either $y = z$ or $y < z$, but y cannot be greater than z. In case 6 either y ______ z or y ______ z, but y cannot be ______ z.

$y = z$ or $y > z$ (either order); less than

---

13 (5) 

(6) 

Cases 5 and 6 can also be represented by symbols. Since, in 5, either y ______ z or y ______ z, we combine these symbols to write $y \leqslant z$ (y is less than or equal to z); and since, in 6, either y ______ z or y ______ z, we combine these symbols to write $y \geqslant z$ (y is greater than or equal to z).

If $y \leqslant z$, then y cannot be ____________ z.
If $y \geqslant z$, then y cannot be ____________ z.
If $y \neq z$, then y cannot be ____________ z.

$y = z$ or $y < z$ (either order); $y = z$ or $y > z$ (either order); greater than; less than; equal to

---

14 Suppose $y = 3$.

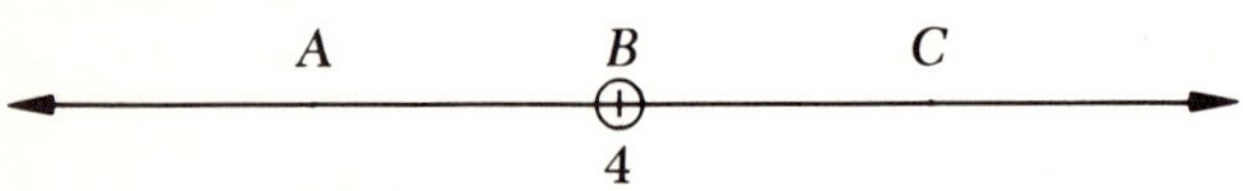

a. Which region is y in? (A/B/C) ______.
b. Which of the following are true statements when $y = 3$?

$y = 4$ $y < 4$ $y > 4$ $y \neq 4$ $y \leqslant 4$ $y \geqslant 4$

---

a. A *b.* $y < 4$; $y \neq 4$; $y \leqslant 4$

---

15 Suppose $y = 5$.

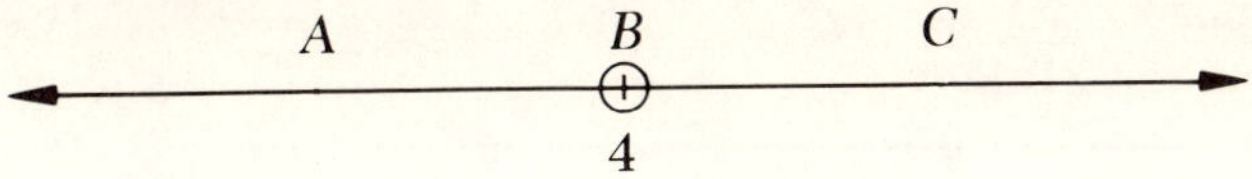

a. Which region is y in? (A/B/C) ______.

b. Which of the following are true statements when $y = 5$?

$y = 4$ $\quad y < 4$ $\quad y > 4$ $\quad y \neq 4$ $\quad y \leqslant 4$ $\quad y \geqslant 4$

---

a. C *b.* $y > 4$; $y \neq 4$; $y \geqslant 4$

---

16 Suppose $y = 4$.

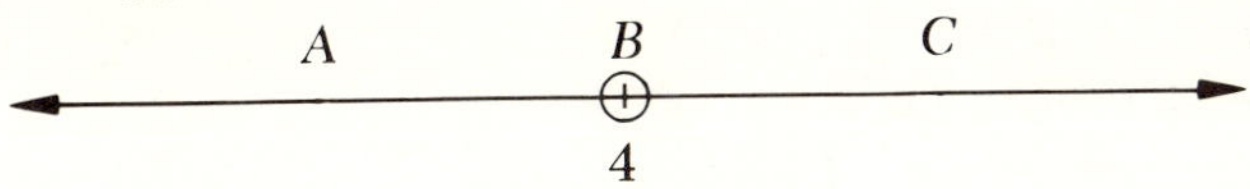

a. Which region is y in? (A/B/C) ______.

b. Which of the following are true statements when $y = 4$?

$y = 4$ $\quad y < 4$ $\quad y > 4$ $\quad y \neq 4$ $\quad y \leqslant 4$ $\quad y \geqslant 4$

---

a. B *b.* $y = 4$; $y \leqslant 4$; $y \geqslant 4$

---

17 Suppose y has two values, 3 and 4.

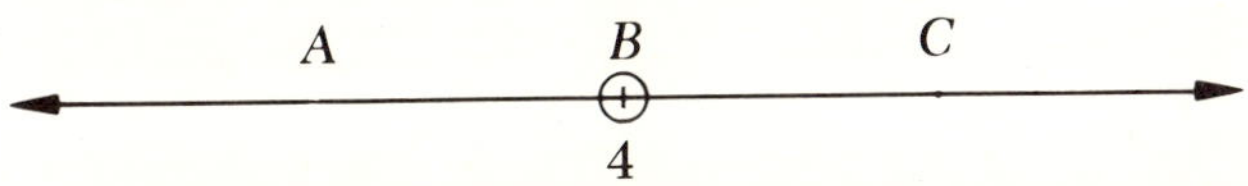

a. Which regions could y be in?

b. Which of the following describes this situation?

$y = 4$ $\quad y < 4$ $\quad y > 4$ $\quad y \neq 4$ $\quad y \leqslant 4$ $\quad y \geqslant 4$

---

a. A and B *b.* $y \leqslant 4$

---

18 Suppose y has two values, 4 and 5.

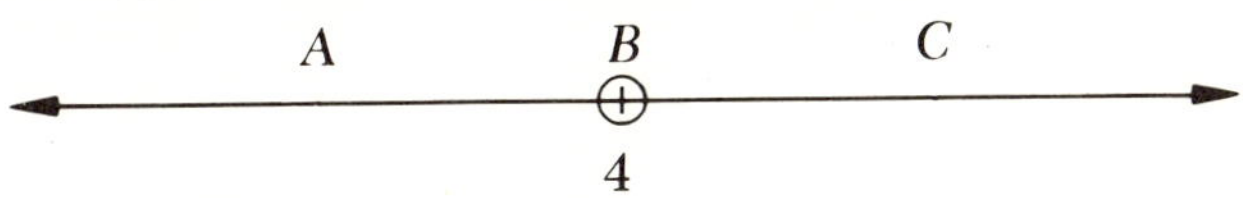

a. Which regions could y be in?

b. Which of the following describes this situation?

$y = 4$ $\quad y < 4$ $\quad y > 4$ $\quad y \neq 4$ $\quad y \leqslant 4$ $\quad y \geqslant 4$

---

a. B and C *b.* $y \geqslant 4$

---

19 Integers can be divided into positive integers, negative integers, and zero.

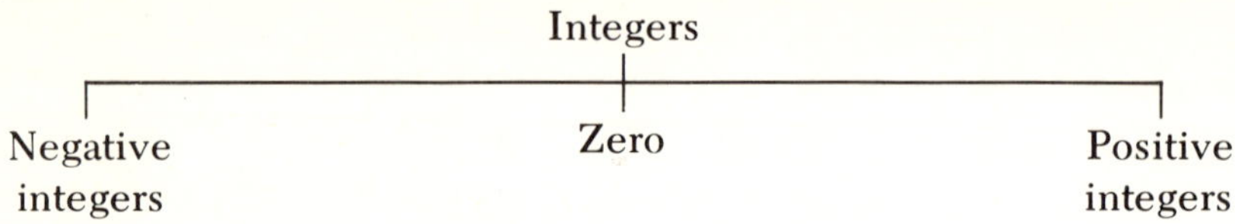

a. If k is a positive integer, then k ($>/\geqslant$) ______ 0.
b. If k is not a negative integer then k ($>/\geqslant$) ______ 0.

---

a. $>$ *b.* $\geqslant$

---

20 *a.* If k is a negative integer, then k ______ 0.
b. If k is not a positive integer, then k ______ 0.

---

a. $k < 0$ *b.* $k \leqslant 0$

---

21 *a.* If $y < 3$, is $y \leqslant 3$? (yes/no) ______
b. If $y = 3$, is $y \leqslant 3$? (yes/no) ______
c. If $y > 3$, is $y \leqslant 3$? (yes/no) ______

---

a. yes *b.* yes *c.* no

---

22 *a.* $8 \geqslant 7$ is (true/false) ______.
b. $7 \geqslant 7$ is (true/false) ______.
c. $6 \geqslant 7$ is (true/false) ______.

---

a. true *b.* true *c.* false

---

23 Which of the following statements is true?

a. $11 \geqslant 10$
b. $4 \leqslant 5$
c. $3 \geqslant 3$
d. $6 \geqslant 7$
e. $1 \leqslant 0$
f. $21 \leqslant 21$

---

a, *b*, *c*, and *f* are true.

---

24 Which of the following statements are true? (If you have any doubt, locate each point on the number line.)

a. $-\frac{3}{2} \geqslant 0$

b. $-4 \leqslant -5$

c. $-3 \leqslant -3$

d. $-\frac{5}{2} \leqslant -\frac{3}{2}$

e. $-\frac{5}{2} \leqslant -\frac{5}{2}$

f. $-6 \geqslant -8$

---

c, *d*, *e*, and *f* are true.

---

25 Suppose that d is less than 10 and greater than 3. Then we know that:

a. d ______ 10 is true.

b. d ______ 3 is true.

c. d ______ 3 can also be written as 3 ______ d.

---

a. $<$ *b.* $>$ *c.* $>$; $<$

---

26 The two statements $3 < d$ and $d < 10$ can be combined into the single statement $3 < d < 10$, which indicates that d is less than ______ and greater than ______. Similarly $5 < g < 8$ indicates that g is less than ______ and greater than ______.

---

10 3 8 5

---

27 Express the following as single statements using $<$ symbols:

a. y is less than 10 and greater than -2.
b. z is less than -3 and greater than -9.

---

a. $-2 < y < 10$ *b.* $-9 < z < -3$

---

28 If k is an integer, what values can k have when:

a. $-11 < k < -5$
b. $-2 < k < 4$

---

a. $-10, -9, -8, -7, -6$ (any order) *b.* $-1, 0, 1, 2, 3$ (any order)

---

29 If k is an integer, what values can k have when:

a. $-3 < k < 2$
b. $-3 \leq k < 2$
c. $-3 \leq k \leq 2$
d. $-3 < k \leq 2$

---

a. $-2, -1, 0, 1$ *b.* $-3, -2, -1, 0, 1$ *c.* $-3, -2, -1, 0, 1, 2$
d. $-2, -1, 0, 1, 2$

---

30 If k is an integer, what values can k have when:

a. $-9 \leq k \leq -6$
b. $0 \leq k < 5$

---

a. $-9, -8, -7, -6$ *b.* $0, 1, 2, 3, 4$

---

31 Express the following as single statements using $<$ or $\leq$ symbols:

a. p is less than 8 and greater than or equal to -3.
b. Z is less than or equal to -4 and greater than -12.

---

a. $-3 \leq p < 8$ *b.* $-12 < Z \leq -4$

---

32 Express the following as single statements using $<$ or $\leq$ symbols:

a. n is less than 0 and greater than or equal to -5.
b. q is less than or equal to -6 and greater than or equal to -13.

---

a. $-5 \leq n < 0$ *b.* $-13 \leq q \leq -6$

---

33 If n is an integer, what values can n have when:

a. $-1 < n \leq 2$
b. $-6 \leq n \leq 0$

---

a. $0, 1, 2$ *b.* $-6, -5, -4, -3, -2, -1, 0$

You have finished the program for Sec. 1-1. Now take the posttest to see how much you have learned.

POSTTEST 1-1

Which of the following statements are true?

1. $5 \neq 4 \times 2$
2. $6 \leqslant 5 + 1$
3. $10 \geqslant 5 \times 2$
4. $8 > 3 + 7$
5. $-8 \neq -4 \times 2$
6. $-6 < -8$
7. If $y = -2$, is $y \geqslant -2$? ______ (yes/no)
8. If $y < 4$, is $y \leqslant 4$? ______ (yes/no)
9. If k is an integer and $0 \leqslant k < 5$, what values can k have?
10. Express the following as a single statement using $<$ or $\leqslant$ signs: n is less than or equal to -3 and greater than -8.

When you have finished the posttest, correct it yourself using the answers on page 57. Then check all the questions you got wrong on the list below:

1. ______	6. ______
2. ______	7. ______
3. ______	8. ______
4. ______	9. ______
5. ______	10. ______

If you compare your performance with that on Pretest 1-1, you will see how much you have improved. If you made more than one mistake on this posttest, review the material and take the pretest again. Don't move on to the next section until you can get all the questions right.

Now proceed to the next section whose number you circled on page 3. If no other sections were circled, take Final Test 1 (page 54).

Section 1-2
The Real Number System

Do this section if you made a mistake in Question 3 of Diagnostic Test 1. Begin by taking the pretest below, which will show you how much of Sec. 1-2 you already know and what you will be able to do by the time you have finished it.

PRETEST 1-2

1. Which of the following numbers are real numbers?
2. Which are rational numbers?
3. Which are irrational numbers?
4. Which are integers?

$$-\sqrt{5} \qquad \sqrt{36} \qquad \frac{0}{-3} \qquad \frac{-8}{2} \qquad \frac{10}{-12} \qquad -6$$

When you have finished the pretest, correct it yourself using the answers on page 57. Then check all the questions you got wrong on the list below:

1. ______ 3. ______

2. ______ 4. ______

If you made any mistakes (even if it was only *one* mistake), read the introduction and go through the program.

If you made no mistakes at all, skip Sec. 1-2, because you already know its material, and go on to the next section whose number you circled on page 3. If no further numbers were circled, take Final Test 1 (page 54).

Introduction

In this section you will have a chance to review the definition of a rational number, and you will meet some irrational numbers, which together with the rational numbers constitute the class of real numbers. You will learn to assign real numbers to the appropriate subclasses. The definition of a real number is omitted because it is very difficult to explain at this stage.

FRAMES FOR SECTION 1-2

1 Which of the following fractions represent rational numbers? Which represent integers?

$\frac{9}{7}$ $\frac{0}{3}$ $\frac{-10}{4}$ $\frac{-6}{-2}$ $\frac{5}{-3}$ $\frac{8}{-4}$

They all represent rational numbers. $\frac{0}{3}$, $\frac{-6}{-2}$, and $\frac{8}{-4}$ represent integers (0, 3, and −2). If you got this right, go on to Frame 6; otherwise continue with Frame 2.

---

2 Any whole number, positive or negative, is called an *integer*. Zero is also an integer. Which of the following represent integers?

2 -3 $\frac{3}{4}$ 0 $-\frac{1}{2}$

---

2 −3 0

---

3 Some fractions represent integers. Which of the following fractions represent integers?

$\frac{2}{2}$ $\frac{3}{4}$ $\frac{-4}{2}$ $\frac{0}{3}$ $\frac{-1}{2}$

$\frac{2}{2}$ $\frac{-4}{2}$ $\frac{0}{3}$

---

4 Any number which can be written in the form $\frac{a}{b}$ in which a and b are both integers and $b \neq 0$ is called a *rational number*. Which of the following represent rational numbers?

$\frac{3}{4}$ $\frac{-6}{6}$ $\frac{3}{1}$ $\frac{-7}{8}$ $\frac{2}{-3}$

---

all of them

---

5 Rational numbers can be divided into two types, integers and nonintegral rational numbers.

Rational numbers

Integers | Nonintegral rational numbers

Which of the following represent integers, and which represent nonintegral rational numbers?

$4 \quad \frac{3}{4} \quad \frac{12}{6} \quad \frac{-8}{2} \quad \frac{-5}{8}$

4, $\frac{12}{6}$, and $\frac{-8}{2}$ represent integers; $\frac{3}{4}$ and $\frac{-5}{8}$ represent nonintegral rational numbers.

---

6 Which of the following represent rational numbers? Which of the following represent integers? Which of the following represent nonintegral rational numbers?

$\frac{7}{9} \quad \frac{0}{5} \quad \frac{-8}{-3} \quad \frac{-9}{-3} \quad \frac{7}{-3} \quad \frac{10}{-5} \quad 5$

All represent rational numbers; $\frac{0}{5}$, $\frac{-9}{-3}$, $\frac{10}{-5}$, and 5 represent integers; $\frac{7}{9}$, $\frac{-8}{3}$, and $\frac{7}{-3}$ represent nonintegral rational numbers.

---

7 In later work we shall find that the class of rational numbers is insufficient for our purposes because we want to use irrational numbers like $\sqrt{2}$ and π. We therefore expand our number system from the rational number system to the real number system.

Real numbers

Rational numbers | Irrational numbers

a. Which of the following are rational numbers?
b. Which of the following are irrational numbers?
c. Which of the following are real numbers?

$\frac{5}{6} \quad \frac{-7}{3} \quad \pi \quad \frac{0}{4} \quad \sqrt{2}$

---

a. $\frac{5}{6}, \frac{-7}{3}$, and $\frac{0}{4}$ are rational numbers. *b.* π and $\sqrt{2}$ are irrational numbers. *c.* They are all real numbers.

---

8 You may have wondered about the radical symbol $\sqrt{\ }$. For positive n, $\sqrt{n}$ is the positive number defined by the equation $\sqrt{n} \times \sqrt{n} = n$. Thus

a. $\sqrt{2} \times \sqrt{2} =$ __________.
b. $(-\sqrt{2}) \times (-\sqrt{2}) =$ __________.

---

a. 2 *b.* 2

---

9 For $n > 0$, $\sqrt{n}$ is defined as positive; if $n = 0$, $\sqrt{n} = 0$. Therefore for $n \geqslant 0$: $\sqrt{n} \times \sqrt{n}$ is (always/sometimes/never) __________ $\geqslant 0$, and $(-\sqrt{n}) \times (-\sqrt{n})$ is (always/sometimes/never) __________ $\geqslant 0$.

---

always; always

---

10 If n is a rational number and $n \geqslant 0$, $\sqrt{n}$ is always a real number and $-\sqrt{n}$ is always a real number. But $\sqrt{-n}$ cannot be defined for the class of real numbers. Which of the following are real numbers?

$\sqrt{\frac{5}{8}}$ $\sqrt{-\frac{6}{7}}$ $-\sqrt{\frac{3}{4}}$

$\sqrt{\frac{5}{8}}$ and $-\sqrt{\frac{3}{4}}$

---

11 *a.* $\sqrt{4} \times \sqrt{4} =$ __________.
b. $2 \times 2 =$ __________.

Therefore

c. $\sqrt{4} =$ __________.

And so

d. $\sqrt{4}$ is a(n) (rational/irrational) __________ number.

---

a. 4 *b.* 4 *c.* $\sqrt{4} = 2$ *d.* rational

---

12 For $n \geq 0$: $\sqrt{n}$ is *always* a(n) (real/rational/irrational) ____________ number, $\sqrt{n}$ is *sometimes* a(n) (real/rational/irrational) ____________ number, for example, $\sqrt{4}$, $\sqrt{9}$, $\sqrt{16}$, and *sometimes* a(n) (real/rational/irrational) ____________ number, for example, $\sqrt{2}$, $\sqrt{3}$, $\sqrt{5}$.

---

real; rational; irrational

---

13 For which two integer values of n does $\sqrt{n} = n$?

---

0 and 1

---

14 The complete real number system is shown in the diagram on page 19; use this diagram to answer questions about the following numbers:

$$\frac{-6}{2} \qquad \sqrt{9} \qquad \frac{3}{-8} \qquad -\sqrt{5} \qquad \frac{0}{-5}$$

a. Which are real numbers?
b. Which are rational numbers?
c. Which are irrational numbers?
d. Which are integers?

---

a. They are all real numbers. *b.* $\frac{-6}{2}$, $\sqrt{9}$, $\frac{3}{-8}$, and $\frac{0}{-5}$ are rational numbers. *c.* $-\sqrt{5}$ is an irrational number. *d.* $\frac{-6}{2}$, $\sqrt{9}$, and $\frac{0}{-5}$ are integers.

---

15 Use the diagram to answer questions about the following numbers:

$$-\sqrt{36} \qquad \frac{-6}{-13} \qquad \frac{12}{-4} \qquad -\sqrt{7} \qquad -2$$

a. Which are real numbers?
b. Which are rational numbers?
c. Which are irrational numbers?
d. Which are integers?

---

a. They are all real numbers. *b.* $-\sqrt{36}$, $\frac{-6}{-13}$, $\frac{12}{-4}$, and -2 are rational numbers. *c.* $-\sqrt{7}$ is an irrational number. *d.* $-\sqrt{36}$, $\frac{12}{-4}$, and -2 are integers.

The Real Number System

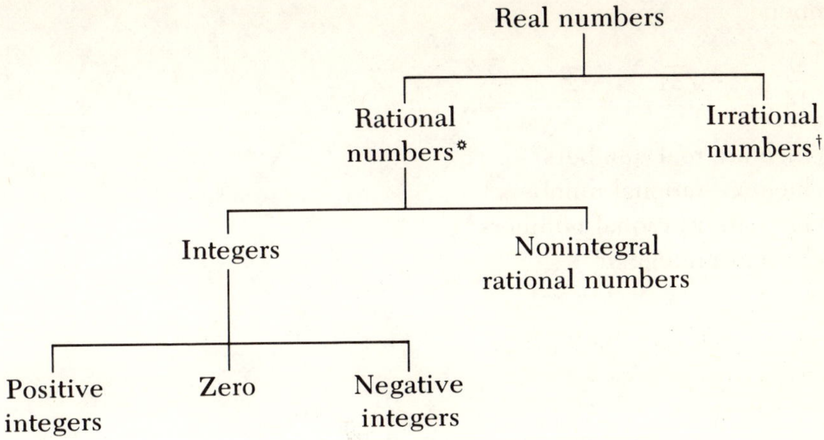

*All numbers which can be written in the form $\frac{a}{b}$, where a and b are integers and $b \neq 0$.

†These include numbers such as $\sqrt{2}$, π, etc., which are real numbers but are not rational numbers.

16 Without referring to the diagram, answer the questions about the following numbers:

$$-\sqrt{16} \qquad \frac{-7}{3} \qquad \frac{-4}{-2} \qquad \frac{0}{4} \qquad \sqrt{3} \qquad -\sqrt{8}$$

a. Which are real numbers?
b. Which are rational numbers?
c. Which are irrational numbers?
d. Which are integers?

---

a. They are all real numbers. *b.* $-\sqrt{16}, \frac{-7}{3}, \frac{-4}{-2}, \frac{0}{4}$ are rational numbers. *c.* $\sqrt{3}$ and $-\sqrt{8}$ are irrational numbers. *d.* $-\sqrt{16}$, $\frac{-4}{-2}$, and $\frac{0}{4}$ are integers.

---

17 Without referring to the diagram, answer the questions about the following numbers:

$\frac{-16}{2}$ $\quad \frac{0}{14}$ $\quad -\sqrt{81}$ $\quad \sqrt{2}$ $\quad \frac{5}{-9}$ $\quad 7$

a. Which are real numbers?
b. Which are rational numbers?
c. Which are irrational numbers?
d. Which are integers?

---

a. They are all real numbers. *b.* 7, $\frac{-16}{2}$, $\frac{0}{14}$, $-\sqrt{81}$, and $\frac{5}{-9}$ are rational numbers. *c.* $\sqrt{2}$ is an irrational number. *d.* 7, $\frac{-16}{2}$, $\frac{0}{14}$, and $-\sqrt{81}$ are integers.

You have finished the program for Sec. 1-2. Now take the posttest to see how much you have learned.

POSTTEST 1-2

1. Which of the following numbers are real numbers?
2. Which are rational numbers?
3. Which are irrational numbers?
4. Which are integers?

$\sqrt{25}$ $\quad -\sqrt{6}$ $\quad \frac{0}{-5}$ $\quad \frac{-10}{2}$ $\quad \frac{12}{-36}$ $\quad -8$

When you have finished the posttest, correct it yourself using the answers on page 57. Then check all the questions you got wrong on the list below:

1. ______ 3. ______

2. ______ 4. ______

If you compare your performance with that on Pretest 1-2, you will see how much you have improved. If you made more than one mistake on this posttest, review the material and take the pretest again. Don't move on to the next section until you can get the questions all right.

Now proceed to the next section whose number you circled on page 3. If no other sections were circled, take Final Test 1 (page 54).

Section 1-3
Notations for Describing Sets

Do this section if you made a mistake in Question 4 or 5 of Diagnostic Test 1. Begin by taking the pretest below, which will show you how much of Sec. 1-3 you already know and what you will be able to do by the time you have finished it.

PRETEST 1-3

1. Which of the following sets is/are identical with $\{3\ 4, 5, 6\}$?

   a. $\{6, 5, 4, 3\}$
   b. $\{3, 4, 5, 4, 6, 4\}$.
   c. $\{3, 4, 5, 7\}$.

Write the following sets in element notation:

2. The set of negative integers
3. $\{n \mid n \text{ is an integer and } n < 3\}$

Write the following sets in property notation:

4. $\{5, 6, 7, 8, \ldots\}$
5. $\{-3, -2, -1, 0, 1, 2, 3\}$

6. Which of the sets in Questions 2 to 5 are infinite?

When you have finished the pretest, correct it yourself using the answers on page 58. Then check all the questions you got wrong on the list below:

1. ______	4. ______
2. ______	5. ______
3. ______	6. ______

If you made any mistakes (even if it was only *one* mistake), read the introduction and go through the program.

If you made no mistakes at all, skip Sec. 1-3, because you already know its material and go on to Sec. 1-4.

Introduction

In this section you meet the concept of a set and learn two ways of representing sets: (1) element notation, which lists the elements of a set, and

(2) property notation, which defines the membership of a set in terms of a common property lacking in nonmembers. You will also learn to change sets of integers from one kind of notation into the other.

FRAMES FOR SECTION 1-3

1 A *set* is a collection of objects defined in such a way that there is no doubt whether a particular object does or does not belong to the set. Which of the following is a set?

a. All old people.
b. All people with birthdays on September 1.

---

b. There are many people whom it would be impossible to categorize either as "old" or "not old."

---

2 If we modified "all old people" to "all people aged 65 or over on January 1, 1971," would it be possible to decide whether or not a person belonged to the group? Could we call the group a set?

---

yes; yes

---

3 A set can have any number of members. If the members of a set can be counted (even if it takes a computer a long time to do it), the set is *finite*. But if trying to count the members would be an endless process (even for a high-speed computer), it is an *infinite* set. Classify the following sets as finite or infinite:

a. The set of all the streets in the city of Chicago.
b. The set of all positive integers.
c. The set of all circles with a particular point as center.
d. The set of all the stitches in a particular curtain.

---

a. finite *b.* infinite *c.* infinite *d.* finite

---

4 If we start counting the members of a set without counting any of them twice, then either we get to a stage when there are no members left to be counted, in which case the set is called __________, or we never get to the end (even after 1 million years) because there are always more members to be counted, in which case the set is called __________.

---

finite; infinite

---

5 The members of a set are also called its *elements*. Two sets are said to be identical if they have the same elements, regardless of the order in which the elements are listed. The set whose elements are the numbers 1, 2, 3, and 4 is written {1, 2, 3, 4}. Which of the following sets is/are identical with it?

a. {4, 3, 2, 1}
b. {1, 2, 3}
c. {2, 1, 3, 4}
d. {1, 2, 3, 4, 5}

---

a and *c*

---

6 Including an element within the braces indicates that that element is a member of the set. It is a waste of space to include the same element twice, because it merely tells us twice that the element is a member of the set.

a. The set {1, 1, 2, 2, 3, 3, 3, 4} has only four elements, the numbers ______, ______, ______, and ______.
b. Which of the following sets is/are identical with it?

{3, 3, 4, 1, 2, 2} {1, 1, 2, 2, 3, 3, 3} {1, 2, 3, 4, 5}

---

a. 1, 2, 3, 4 *b.* {3, 3, 4, 1, 2, 2}

---

7 One method of notation used by mathematicians in referring to sets (the one we have just begun to use) is called *element notation* because it lists all the elements of the set and encloses them within braces. Thus the set whose elements are the numbers 5, 6, and 7 is written {5, 6, 7}. How should we write the set of all positive integers greater than 6 and less than 12?

{7, 8, 9, 10, 11}

---

8 Use element notation to write:

a. The set of integers greater than 2 and less than 9.
b. The set of integers greater than 0 and less than 4.

---

a. {3, 4, 5, 6, 7, 8} *b.* {1, 2, 3}

---

9 We can also use element notation for infinite sets if we use three dots in order to indicate that a sequence of elements goes on forever. Thus the set of integers greater than 1 can be written $\{2, 3, 4, 5, 6, \ldots\}$. Notice that 1 (is/is not) ______ a member of this set. How would you write the set of integers greater than 4?

---

is not; $\{5, 6, 7, 8, \ldots\}$

---

10 Use element notation to write:

a. The set of integers greater than 6.
b. The set of integers greater than -1.

---

a. $\{7, 8, 9, 10, 11, \ldots\}$ *b.* $\{0, 1, 2, 3, 4, \ldots\}$

---

11 The set of integers less than 4 is written $\{\ldots, -3, -2, -1, 0, 1, 2, 3\}$ How would you write the set of integers less than 2?

$\{\ldots, -4, -3, -2, -1, 0, 1\}$

---

12 Use element notation to write:

a. The set of integers less than 3.
b. The set of integers less than -2.
c. The set of integers greater than -3.

---

a. $\{\ldots, -3, -2, -1, 0, 1, 2\}$ *b.* $\{\ldots, -6, -5, -4, -3\}$
c. $\{-2, -1, 0, 1, 2, 3, \ldots\}$

---

13 Use element notation to write:

a. The set of positive integers.
b. The set of negative integers.

---

a. $\{1, 2, 3, 4, 5, \ldots\}$ *b.* $\{\ldots, -5, -4, -3, -2, -1\}$

---

14 Another notation, called *property notation*, uses a property to define a set: any object which has the property is an element of the set. In this notation the set of rational numbers is written as $\{y \mid y \text{ is a rational number}\}$ because y is in the set if and only if it is a rational number. How should we write the set of real numbers in property notation?

$\{y \mid y \text{ is a real number}\}$

---

15 It does not matter which letter you use. The answer to Frame 14 could also have been written $\{b \mid b$ is a real number$\}$ or $\{g \mid g$ is a real number$\}$. Write the set of integers in property notation, using the letter n.

$\{n \mid n$ is an integer$\}$

---

16 $\{7, 8, 9, 10, \ldots\}$ can be described in property notation as $\{d \mid d$ is an integer and $d > 6\}$.

a. $\{3, 4, 5, 6, \ldots\}$ can be described as $\{z \mid z$ is an integer and $z >$ _____$\}$.
b. $\{11, 12, 13, 14, \ldots\}$ can be described as $\{y \mid y$ is an integer and _____$\}$.

---

a. 2 *b.* $y > 10$

---

17 *a.* $\{-3, -2, -1, 0, \ldots\}$ can be described as $\{p \mid p$ is an integer and $p >$ _____$\}$.
b. $\{-6, -5, -4, -3, \ldots\}$ can be described as $\{b \mid b$ is an integer and _____$\}$.

---

a. -4 *b.* $b > -7$

---

18 *a.* $\{\ldots, -1, 0, 1, 2\}$ can be described as $\{z \mid z$ is an integer and $z <$ _____$\}$.
b. $\{\ldots -9, -8, -7, -6\}$ can be described as $\{d \mid d$ is an integer and _____$\}$.

---

a. 3 *b.* $d < -5$

---

19 Describe the following sets in property notation:

a. $\{3, 4, 5, 6, \ldots\}$
b. $\{-2, -1, 0, 1, \ldots\}$
c. $\{\ldots, -6, -5, -4, -3\}$
d. $\{\ldots, 3, 4, 5, 6\}$

---

a. $\{y \mid y$ is an integer and $y > 2\}$
b. $\{y \mid y$ is an integer and $y > -3\}$
c. $\{y \mid y$ is an integer and $y < -2\}$
d. $\{y \mid y$ is an integer and $y < 7\}$

---

20 Conversely, $\{y \mid y$ is an integer and $y > 2\}$ can be described in element notation as $\{$____, ____, ____, ____, . . .$\}$, and $\{q \mid q$ is an integer and $q < -3\}$ can be described in element notation as ________________.

$\{3, 4, 5, 6, \ldots\}$ $\{\ldots, -7, -6, -5, -4\}$

---

21 Describe the following sets in element notation:

a. $\{g \mid g$ is an integer and $g > -1\}$
b. $\{f \mid f$ is an integer and $f < 2\}$
c. $\{d \mid d$ is an integer and $d < -5\}$

---

a. $\{0, 1, 2, 3, \ldots\}$ *b.* $\{\ldots, -2, -1, 0, 1\}$
c. $\{\ldots, -9, -8, -7, -6\}$

---

22 Were any of the sets in Frame 21 finite?

No, you could go on counting their members forever and still not finish.

---

23 $\{7, 8, 9, 10, 11\}$ can be described in property notation as $\{d \mid d$ is an integer and $6 < d < 12\}$.

a. $\{12, 13, 14, 15\}$ can be described as $\{g \mid g$ is an integer and _____ $< g <$ _____$\}$.
b. $\{2, 3, 4, 5, 6, 7, 8, 9\}$ can be described as $\{h \mid h$ is an integer and _____$\}$.

---

a. 11; 16 *b.* $1 < h < 10$

---

24 *a.* $\{-4, -3, -2, -1, 0, 1, 2\}$ can be described as $\{y \mid y$ is an integer and _____ $< y <$ _____$\}$.
b. $\{-8, -7, -6, -5, -4\}$ can be described as $\{z \mid z$ is an integer and __________$\}$.

---

a. -5; 3 *b.* $-9 < z < -3$

---

25 Describe the following in property notation:

a. $\{-2, -1, 0, 1, 2, 3, 4, 5\}$
b. $\{-7, -6, -5, -4, -3, -2\}$

---

a. $\{y \mid y$ is an integer and $-3 < y < 6\}$ *b.* $\{z \mid z$ is an integer and $-8 < z < -1\}$

---

26 Describe the following in element notation:

a. $\{g \mid g$ is an integer and $-5 < g < 3\}$
b. $\{h \mid h$ is an integer and $-11 < h < -3\}$

---

a. $\{-4, -3, -2, -1, 0, 1, 2\}$ *b.* $\{-10, -9, -8, -7, -6, -5, -4\}$

You have finished the program for Sec. 1-3. Now take the posttest to see how much you have learned.

POSTTEST 1-3

1. Which of the following sets is/are identical with $\{0, 1, 2, 3\}$?

   a. $\{3, 2, 1, 0\}$
   b. $\{0, 1, 2, 1, 3, 1\}$
   c. $\{0, 1, 2, 4\}$

Write the following sets in element notation:

2. The set of positive integers

3. $\{n \mid n$ is an integer and $n > -2\}$

Write the following sets in property notation:

4. $\{3, 4, 5, 6, \ldots\}$

5. $\{-5, -4, -3, -2, -1, 0, 1\}$

6. Which of the sets in Questions 2 to 5 are finite?

When you have finished the posttest, correct it yourself using the answers on page 58. Then check all the questions you got wrong on the list below:

1. ______ 4. ______

2. ______ 5. ______

3. ______ 6. ______

If you compare your performance with that on Pretest 1-3, you will see how much you have improved. If you made more than one mistake on this posttest, review the material and take the pretest again. Don't move on to the next section until you can get all the questions right. Now proceed to Sec. 1-4.

Section 1-4
Intersection and Union of Sets

Do this section if you made a mistake in Question 4 or 5 of Diagnostic Test 1. Begin by taking the pretest below, which will show you how much of Sec. 1-4 you already know and what you will be able to do by the time you have finished it.

PRETEST 1-4

1. Is $\{-1, 0, 1\}$ a subset of $\{k \mid k$ is an integer and $k > -1\}$?

2. Is $\{q \mid q$ is a real number$\}$ a subset of $\{z \mid z$ is a rational number$\}$?

Write the following sets in element notation:

3. $\{-5, -4, 4, 5\} \cup \{1, 2, 3, 4\}$

4. $\{-4, -3, -2, -1, 0\} \cap \{-1, 0, 1, 2, 3\}$

Write the following sets in property notation:

5. $\{a \mid a$ is a rational number$\} \cap \{b \mid b$ is an irrational number$\}$

6. $\{c \mid c$ is a rational number$\} \cup \{d\,]\, d$ is an integer$\}$

7. $\{e \mid e$ is an irrational number$\} \cap \{f \mid f$ is a real number$\}$

When you have finished the pretest, correct it yourself using the answers on page 58. Then check all the questions you got wrong on the list below:

1. ______ 5. ______
2. ______ 6. ______
3. ______ 7. ______
4. ______

If you made any mistakes (even if it was only *one* mistake), read the introduction and go through the program.

If you made no mistakes at all, skip Sec. 1-4, because you already know its material, and go on to the next section whose number you circled on page 3. If no further numbers were circled, take Final Test 1 (page 54).

Introduction

In this section you will meet the concepts of the empty set, of the union and intersection of sets, and of a subset; and you will learn to use the symbols $\emptyset$, $\cup$, and $\cap$. You will then practice these concepts and the use of set notation.

FRAMES FOR SECTION 1-4

1 How many elements do the following sets have?

a. $\{1, 1, 2, 2, 3, 3, 4\}$
b. $\{2\}$
c. $\{a \mid a$ is an integer and $2 < a < 4\}$
d. $\{b \mid b$ is an integer and $2 < b < 3\}$

---

a. 4 *b.* 1 *c.* 1 *d.* 0

---

2 How many elements do the following sets have?

a. $\{-5, -4, -4, -2\}$
b. $\{p \mid p$ is an integer and $-5 < p < -2\}$
c. $\{q \mid q$ is an integer and $-3 < q < -2\}$
d. $\{r \mid r$ is an integer and $-4 < r < -2\}$

---

a. 3 *b.* 2 *c.* 0 *d.* 1

---

3 In Sec. 1-3 we called two sets identical if they had the same elements. Thus all sets which have 1 and 2 as elements but no other elements are identical with the set $\{1, 2\}$. Similarly all sets which have no elements at all are identical. We call the set with no elements the *empty set* and denote it by $\emptyset$ (no braces). Rewrite the following sets in element notation, using $\emptyset$ for those sets which are identical with the empty set.

a. $\{p \mid p$ is an integer and $2 < p < 4\}$
b. $\{p \mid p$ is an integer and $2 < p < 3\}$
c. $\{k \mid k$ is a man with four eyes$\}$
d. $\{m \mid m$ was a female president of the United States$\}$

---

a. $\{3\}$ *b.* $\emptyset$ *c.* $\emptyset$ *d.* $\emptyset$

---

4 Rewrite the following sets in element notation:

a. $\{d \mid d \text{ is an integer and } -3 < d < 0\}$
b. $\{f \mid f \text{ is a positive integer and } f < -2\}$
c. $\{n \mid n \text{ is a man over 200 years old}\}$

---

a. $\{-2, -1\}$ *b.* $\emptyset$ *c.* $\emptyset$

---

5 Rewrite the following sets in element notation:

a. $\{d \mid d \text{ is a dog with 16 legs}\}$
b. $\{g \mid g \text{ is an integer and } -2 < g < 3\}$
c. $\{h \mid h \text{ is a positive rational number and } h < 0\}$

---

a. $\emptyset$ *b.* $\{-1, 0, 1, 2\}$ *c.* $\emptyset$

---

6 The *union* of sets A and B (written $A \cup B$) is the set which contains all the elements of A and all the elements of B but no other element. Thus an element is in the set $A \cup B$ if it is in *either A or B or both.* If A is $\{1, 2, 3, 4\}$ and B is $\{5, 6, 7, 8\}$, then the set $A \cup B$ is ________.

$\{1, 2, 3, 4, 5, 6, 7, 8\}$

---

7 What is the shortest way of writing $A \cup B$ if A is $\{0, 1, 2, 3\}$ and B is $\{2, 3, 4, 5\}$? (Remember that there is no point in listing an element more than once.)

$\{0, 1, 2, 3, 4, 5\}$

---

8 Write the following sets in the shortest way:

a. $\{-2, -1, 0, 1, 2, 3, 4\} \cup \{1, 2, 3, 4, 5, 6\}$
b. $\{-3, -2, -1\} \cup \{2, 3, 4\}$
c. $\{-7, -6, -5, -4, -3, -2\} \cup \{-6, -5, -4\}$

---

a. $\{-2, -1, 0, 1, 2, 3, 4, 5, 6\}$ *b.* $\{-3, -2, -1, 2, 3, 4\}$
c. $\{-7, -6, -5, -4, -3, -2\}$

---

9 The area common to two highways (see diagram) is called the *intersection* of the two highways.

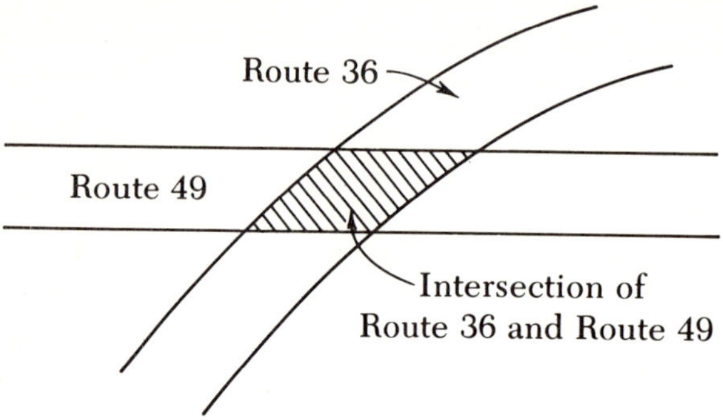

The intersection of sets A and B is defined in a similar manner. It contains all the elements which are common to both sets and no others. It is written $A \cap B$. Thus an element will be in the set $A \cap B$ only if it is in *both A and B*. If A is {1, 2, 3, 4} and B is {3, 4, 5, 6}, what set is $A \cap B$?

{3, 4}

---

10 You can remember the difference between the union symbol ∪ and the intersection symbol ∩ by the fact that ∪, which stands for union, looks like a letter U.

a. If A is $\{-2, -1, 0, 1, 2\}$ and B is {0, 1, 2, 3, 4}, what set is $A \cap B$?
b. What set is $A \cup B$?

---

a. {0, 1, 2} *b.* $\{-2, -1, 0, 1, 2, 3, 4\}$

---

11 If A is $\{-7, -6, -5, 5, 6, 7\}$ and B is $\{-6, -5, -4, -3, -2, -1\}$, what set is:

a. $A \cup B$
b. $A \cap B$

---

a. $\{-7, -6, -5, -4, -3, -2, -1, 5, 6, 7\}$ *b.* $\{-6, -5\}$

---

12 Insert the symbol ∪ or ∩ to make the following statements true:

a. {5, 6, 7} = {4, 5, 6, 7} ______ {5, 6, 7, 8, 9}
b. {4, 5, 6, 7, 8, 9} = {4, 5, 6, 7} ______ {5, 6, 7, 8, 9}
c. $\{-5, -4, -3, -2, -1\} = \{-5, -4, -3\}$ ______ $\{-4, -3, -2, -1\}$
d. $\{-4, -3\} = \{-5, -4, -3\}$ ______ $\{-4, -3, -2, -1\}$

a. ∩ *b.* ∪ *c.* ∪ *d.* ∩

13 Insert the symbol ∪ or ∩ to make the following statements true:

a. $\{1, 2, 3, 4\} = \{1, 2, 3\}$ ______ $\{2, 3, 4\}$
b. $\{2, 3\} = \{1, 2, 3\}$ ______ $\{2, 3, 4\}$
c. $\emptyset = \{1, 2\}$ ______ $\{3, 4\}$
d. $\{1, 2, 3, 4\} = \{1, 2\}$ ______ $\{3, 4\}$

a. ∪ *b.* ∩ *c.* ∩ *d.* ∪

14 Insert the symbol ∪ or ∩ to make the following statements true:

a. $\{-4, -3, -2, -1, 0, 1, 2\} = \{-4, -3, -2, -1, 0\}$ ______ $\{0, 1, 2\}$
b. $\{-1, 0, 1, 2\} = \{-1, 0\}$ ______ $\{0, 1, 2\}$
c. $\emptyset = \{-2, -1\}$ ______ $\{0, 1\}$
d. $\{0\} = \{-1, 0\}$ ______ $\{0, 1\}$

a. ∪ *b.* ∪ *c.* ∩ *d.* ∩

15 Write the following sets in element notation:

a. $\{-2, -1, 0, 1, 2, 3, 4\} \cap \{1, 2, 3, 4, 5, 6, 7\}$
b. $\{-3, -2, -1\} \cap \{2, 3, 4\}$
c. $\{-3, -2, -1, 0\} \cup \{5, 6, 7, 8\}$

a. $\{1, 2, 3, 4\}$ *b.* $\emptyset$ *c.* $\{-3, -2, -1, 0, 5, 6, 7, 8\}$

16 Write the following sets in element notation:

a. $\{-6, -5, -4\} \cap \{-3, -2, -1\}$
b. $\{0, 1, 2, 3, 4\} \cup \{1, 2, 3, 4, 5\}$
c. $\{-3, -2, -1, 0\} \cap \{0, 1, 2, 3\}$

a. $\emptyset$ *b.* $\{0, 1, 2, 3, 4, 5\}$ *c.* $\{0\}$

17 If A is the set of all points on a given straight line and B is the set of all points on another straight line in the same plane, how many elements has $A \cap B$ when the two lines are (*a*) parallel, (*b*) nonparallel?

c. What set is $A \cap B$ when the lines are parallel?

a. 0 *b.* 1 *c.* $\emptyset$

18 If every element in set A is also an element in set B, then A is called a *subset* of B. Thus $\{2, 3, 4\}$ (is/is not) ________ a subset of $\{1, 2, 3, 4\}$, but $\{2, 3, 5\}$ (is/is not) ________ a subset of $\{1, 2, 3, 4\}$ because the element ________ is not a member of $\{1, 2, 3, 4\}$.

is; is not; 5

19 *a.* Is $\{1, 3, 5\}$ a subset of $\{1, 2, 3, 4, 5\}$?
b. Is $\{-2, -3\}$ a subset of $\{1, 2, 3, 4, 5\}$?
c. Is $\{3, 4, 5, 6\}$ a subset of $\{n \mid n \text{ is an integer}\}$?

a. yes *b.* no *c.* yes

20 In deciding whether A is a subset of B, carefully consider each element of A to see if it is also an element of B.

a. Is $\{3, 4, 5\}$ a subset of $\{k \mid k \text{ is an integer and } k > 4\}$?
b. Is $\{3, 4, 5\}$ a subset of $\{k \mid k \text{ is an integer and } k > 1\}$?
c. Is $\{3, 4, 5\}$ a subset of $\{k \mid k \text{ is an integer and } k > 3\}$?

a. no *b.* yes *c.* no

21 *a.* Is $\{-4, -3, -2\}$ a subset of $\{k \mid k \text{ is an integer and } k < -1\}$?
b. Is $\{-3, -2, -1\}$ a subset of $\{h \mid h \text{ is an integer and } h < -1\}$?
c. Is $\{-2, -1, 0\}$ a subset of $\{m \mid m \text{ is an integer and } m < -1\}$?

a. yes *b.* no *c.* no

22 Refer to the diagram on page 19 if you need help.

a. Is $\{n \mid n \text{ is an integer}\}$ a subset of $\{k \mid k \text{ is a rational number}\}$?
b. Is $\{y \mid y \text{ is a real number}\}$ a subset of $\{m \mid m \text{ is a rational number}\}$?
c. Is $\{z \mid z \text{ is a rational number}\}$ a subset of $\{q \mid q \text{ is a real number}\}$?

a. yes *b.* no *c.* yes

23 The diagrams below depict three possible kinds of situation.

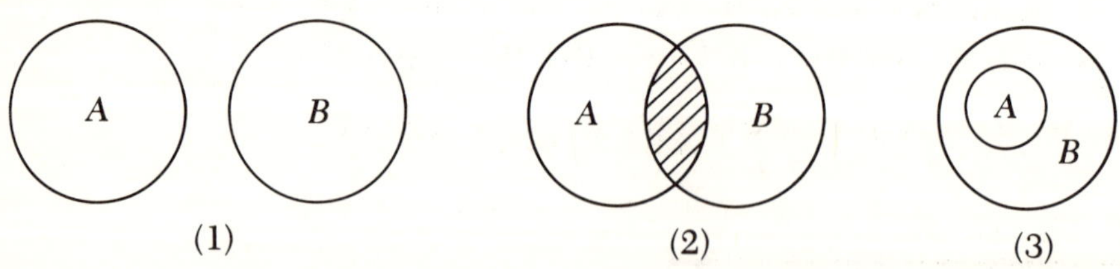

A is the set of all points in one circle, and B is the set of all points in the other circle.

a. In situation 1, $A \cap B =$ ______.
b. In situation 2, the shaded area is A ______ B ($\cup/\cap$).
c. In situation 3, A is a ______ of B.

---

a. $\emptyset$ *b.* $\cap$ *c.* subset

---

24 The shaded area in each diagram corresponds to $A \cup B$ or $A \cap B$. Specify which.

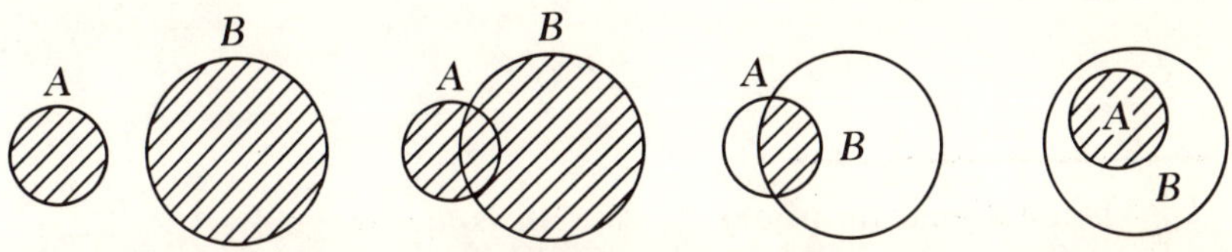

a. *b.* *c.* *d.*

---

a. $A \cup B$ *b.* $A \cup B$ *c.* $A \cap B$ *d.* $A \cap B$

---

25 The shaded area in each diagram corresponds to $A \cup B$ or $A \cap B$. Specify which.

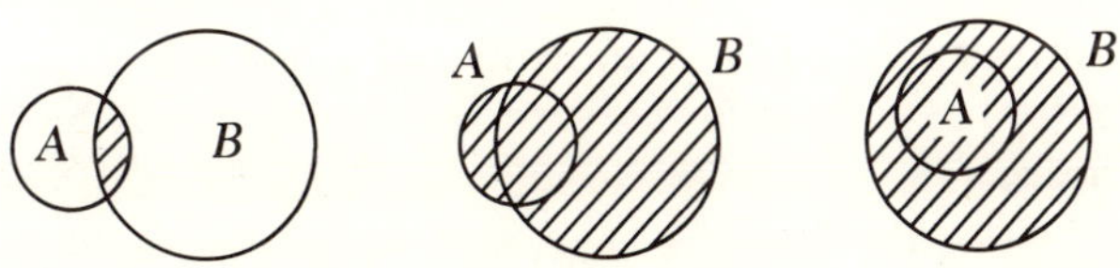

a. *b.* *c.*

---

a. $A \cap B$ *b.* $A \cup B$ *c.* $A \cup B$

---

26 Looking back at Frames 24 and 25, you should notice that if A is a subset of B, then:

a. $A \cap B =$ ______
b. $A \cup B =$ ______

---

a. A *b.* B

---

27 If C is a subset of D then:

a. $C \cup D =$ ______
b. $C \cap D =$ ______

---

a. D *b.* C

---

28 *a.* If B is the set $\{b \mid b$ is an integer$\}$ and D is the set $\{d \mid d$ is a positive integer$\}$, then __________ is a subset of __________.
b. $B \cup D =$ __________.
c. $\{b \mid b$ is an integer$\} \cap \{d \mid d$ is a positive integer$\} =$ __________.

---

a. D is a subset of B. *b.* B or $\{b \mid b$ is an integer$\}$
c. D or $\{d \mid d$ is a positive integer$\}$

---

29 *a.* $\{k \mid k$ is a rational number$\} \cup \{m \mid m$ is an integer$\} =$ ______.
b. $\{k \mid k$ is a rational number$\} \cap \{m \mid m$ is an integer$\} =$ ______.

---

a. $\{k \mid k$ is a rational number$\}$ *b.* $\{m \mid m$ is an integer$\}$

---

30 (Refer back to the diagram on page 19 if you need help.)

a. $\{d \mid d$ is a rational number$\} \cup \{e \mid e$ is a real number$\} =$ __________.

b. $\{d \mid d$ is a rational number$\} \cap \{f \mid f$ is an irrational number$\} =$ __________.

---

a. $\{e \mid e$ is a real number$\}$ *b.* $\emptyset$

---

31 *a.* $\{d \mid d$ is a rational number$\} \cup \{f \mid f$ is an irrational number$\} =$ __________.

b. $\{f \mid f$ is an irrational number$\} \cap \{e \mid e$ is a real number$\} =$ __________.

c. $\{a \mid a$ is a positive integer$\} \cap \{b \mid b$ is a negative integer$\} =$ __________.

---

a. $\{e \mid e$ is a real number$\}$ *b.* $\{f \mid f$ is an irrational number$\}$
c. $\emptyset$

You have finished the program for Sec. 1-4. Now take the posttest to see how much you have learned.

POSTTEST 1-4

1. Is $\{-3, -2, -1\}$ a subset of $\{k \mid k \text{ is an integer and } k < -1\}$?

2. Is $\{n \mid n \text{ is an integer}\}$ a subset of $\{m \mid m \text{ is a rational number}\}$?

Write the following sets in element notation:

3. $\{-5, -4, 4, 5\} \cap \{1, 2, 3, 4\}$

4. $\{-4, -3, -2, -1, 0\} \cup \{-1, 0, 1, 2, 3\}$

Write the following sets in property notation:

5. $\{a \mid a \text{ is a positive integer}\} \cap \{b \mid b \text{ is a negative integer}\}$

6. $\{c \mid c \text{ is a positive integer}\} \cup \{d \mid d \text{ is an integer}\}$

7. $\{e \mid e \text{ is a rational number}\} \cap \{f \mid f \text{ is a real number}\}$

When you have finished the posttest, correct it yourself using the answers on page 58. Then check all the questions you got wrong on the list below:

1. ______ 5. ______

2. ______ 6. ______

3. ______ 7. ______

4. ______

If you compare your performance with that on pretest 1-4, you will see how much you have improved. If you made more than one mistake on this posttest, review the material and take the pretest again. Don't move on to the next section until you can get all the questions right.

Now proceed to the next section whose number you circled on page 3. If no other sections were circled, take Final Test 1 (page 54).

Section 1-5
Variables and Domains, Open Sentences, and Closed Sentences

Do this section if you made a mistake in Questions 6 to 8 of Diagnostic Test 1. Begin by taking the pretest below, which will show you how much of Sec. 1-5 you already know and what you will be able to do by the time you have finished it.

PRETEST 1-5

1. Which of the following expressions contains a variable?

   a. $3z + 5$, where z is a rational number
   b. $(3 \times 2) + (5 \times 6)$
   c. $(5 \times y) + (3 \times y)$, where y is 1, 2, or 3

2. What are the domains of the variables of Question 1?

3. Label the following sentences as equations, equalities, inequations, or inequalities.

   a. $7 + 3 \neq 11$ *b.* $8z > 5$
   c. $3y = 9$ *d.* $5k \neq 11$
   e. $7 + 3 = 11$ *f.* $-4 < -1$

4. Which of the sentences in Question 3 are open sentences, and which are closed sentences? Which are true and which false?

Which of the following sentences are open, which are true, and which are false?

5. $q + 3 = q + 3$ 6. $k - 5 \leq k - 6$

7. $4w = 16$ when the domain of w is $\{1, 2, 3, 4\}$

8. $5m \geq 15$ when the domain of m is $\{3, 4, 5, \ldots\}$

When you have finished the pretest, correct it yourself using the answers on pages 58 and 59. Then check all the questions you got wrong on the list below:

1. ______ 5. ______

2. ______ 6. ______

3. ______ 7. ______

4. ______ 8. ______

If you made any mistakes (even if it was only *one* mistake), read the introduction and go through the program.

If you made no mistakes at all, skip Sec. 1-5, because you already know its material, and go on to the next section whose number you circled on page 3. If no further numbers were circled, take Final Test 1 (page 54).

Introduction

After introducing the concept of a variable and its domain, this section explains the terms *expression* and *sentence*. It then shows you how to classify sentences as equations, equalities, inequations, or inequalities, and as open, closed, true, or false.

FRAMES FOR SECTION 1-5

1 In algebra we often use letters to summarize a lot of information in a single statement. Consider, for example, the sentence $n \times 0 = 0$, where n is any positive integer. Would it be easy to write this information in a collection of statements like $1 \times 0 = 0$, $2 \times 0 = 0$, etc.? Explain your answer.

No; we would have to write an infinite number of statements like $1 \times 0 = 0$ in order to convey the same amount of information, and so we would never stop writing!

---

2 *a.* $n \times 0 = 0$, where n is any positive integer.
b. $d \times 0 = 0$, where d is any negative integer.
c. $p \times 0 = 0$, where p is any rational number.

In sentence *a*, n is called a *variable over the set of positive integers*. In sentence *b*, d is called a ___ ___ ___ ___ ___ ___ ___. In sentence *c*, p is called a ___ ___ ___ ___ ___ ___ ___.

---

variable over the set of negative integers
variable over the set of rational numbers

---

3 The set whose elements consist of those numbers which a variable can represent is called the *domain* of that variable. Thus in Frame 2*a*, the domain of the variable n is the set of positive integers. What is the domain of the variable p in Frame 2*c*? What is the domain of the variable p in this sentence: $p > 3$ if p is an element of the set $\{4, 5, 6\}$?

---

the set of rational numbers; $\{4, 5, 6\}$

---

4 If s is a variable over the set of numbers $\{1, 2, 3, 4, 5\}$, then the domain of s is {____________}. If t is 5, 6, or 7, the domain of t is {____________}.

$\{1, 2, 3, 4, 5\}$ $\{5, 6, 7\}$

---

5 If w is 2, 4, 6, or 8, the domain of w is {____________}.

$\{2, 4, 6, 8\}$

---

6 In Frame 2*b* the set of negative integers is the ____________ of the ____________ *d*.

---

domain; variable

---

7 *a.* $k \times 0 = 0$, where k is any positive integer.
b. $k \times 0 = 0$, where k is any rational number.
c. $k \times 0 = 0$, where k is any real number.

Do these three sentences give the same information? If so, what information? If not, what is different about the information given by each sentence.

No; *a* gives information only about positive integers; *b* gives information only about rational numbers; and *c* gives information about all real numbers.

---

8 In order to know what information is given in a sentence containing a variable, you need to know the ____________ of that variable.

---

domain

---

9 If the domain of the variable is not given, it is usually assumed to be the set of real numbers. Thus in the sentence $\frac{y}{y} = 1$, the domain of y is assumed to include which of the following numbers:

a. 6
b. $-3\frac{3}{4}$
c. $\sqrt{2}$
d. $-\sqrt{3}$

Refer to the diagram on page 19 if you need to.

---

all of them

---

10 Unless the domain of a variable is specifically stated, it is assumed to be ____ ____ ____ ____ ____ ____.

---

the set of all real numbers

---

11 The building blocks of algebra are *expressions*. Expressions contain combinations of numbers, variables, and operation signs, for example, +, −, ×, ÷, and $\sqrt{\ }$. The following are all expressions. Which of them contain variables?

a. $\left\{2 + 6\left(y - \frac{3}{4}\right)\right\}$

b. 5

c. $\frac{7}{13} + 3(z - 2)$

d. $2p - 1$, where p is an integer

e. t, where t is 1, 2, or 3

---

a, *c*, *d*, and *e*

---

12 What are the domains of the variables in expressions *c*, *d*, and *e* of Frame 11?

---

c. the set of real numbers *d.* the set of integers *e.* $\{1, 2, 3\}$

---

13 Any two expressions can be related to each other by the use of one of the following six symbols:

$=$ meaning "equal to"
$\neq$ meaning "not equal to"
$<$ meaning "________ than"
$>$ meaning "________ than"
$\leqslant$ meaning "____ ____ ____ ____ ____"
$\geqslant$ meaning "____ ____ ____ ____ ____"

---

less; greater; less than or equal to; greater than or equal to

---

14 Two expressions linked by one of these six symbols form a *sentence*. Sentences can be classified into *equalities*, *equations*, *inequalities*, and *inequations* by the table below.

Linking symbol / Presence of a variable	$<, \leqslant, >, \geqslant, \neq$	$=$
Variable present	inequation	equation
No variable present	inequality	equality

Label the following sentences as equation, equality, inequation, or inequality:

a. $7 \leqslant 8$
b. $5 = z + 7$
c. $6(7 + 5) < 11y$
d. $11 \neq 10 + 5$
e. $13p + 1 \geqslant 6(7p + 2)$
f. $10 = 6 + 4$

---

a. inequality *b.* equation *c.* inequation *d.* inequality
e. inequation *f.* equality

---

15 Label the following sentences as equation, equality, inequation, or inequality:

a. $-6 \neq 0$

b. $3(z + 5) = 21$

c. $\frac{y}{3} + 2 = 7$

d. $p \neq 8$

e. $-2 < d + 4$

f. $10 > 8 + 4$

---

a. inequality *b.* equation *c.* equation *d.* inequation
e. inequation *f.* inequality

---

16 If $y = 2$, is sentence *c* in Frame 15 true? If $y = 15$, is sentence *c* in Frame 15 true? If $z = 2$, is sentence *b* in Frame 15 true? If $z = 15$, is sentence *b* in Frame 15 true? Can you tell whether sentences *b*, *c*, *d*, and *e* are true or false without knowing the value of the variable?

---

no; yes; yes; no; no

---

17 If you can tell whether a sentence is true or false, it is called a *closed sentence*. If you cannot tell whether a sentence is true or false it is called an *open sentence*. Which of the following sentences are closed, and which are open?

a. $-5 \neq 3$

b. $3(z + 5) = 21$

c. $\dfrac{y}{3} + 2 = 7$

d. $10 > 3 + 14$

e. $-2 < d + 4$

f. $2 \times 4 = 7$

Which of the closed sentences are true?

---

a, *d*, and *f* are closed sentences; *b*, *c*, and *e* are open sentences; only *a* is a true sentence.

---

18 An inequality may be true, as in Frame 17*a* or false, as in Frame 17*d*, but it must always be (open/closed) ______; and an equality may be true, as in $2 + 2 = 4$, or false, as in Frame 17*f*, but it also must always be (open/closed) ______.

---

closed; closed

---

19 Consider the sentence $p > 4$.

a. If the domain of p is $\{1, 2, 3\}$, the sentence is (true/false/open) ______.
b. If the domain of p is $\{5, 6, 7\}$, the sentence is (true/false/open) ______.
c. If the domain of p is $\{1, 2, 3, 4, 5, 6, 7\}$ the sentence is (true/false/open) ______.

---

a. false *b.* true *c.* open

---

20 Consider the sentence $-2 \leqslant d + 4$.

a. If the domain of d is the set of positive integers, the sentence is (true/false/open) ________.
b. If the domain of d is the set of negative integers, the sentence is (true/false/open) ________.

---

a. true *b.* open

---

21 Complete the sentences below with one of the following: true, false, open, true or false, true or false or open:

a. An equality is ________.
b. An inequality is ________.
c. An equation is ________.
d. An inequation is ________.

---

a. true or false *b.* true or false *c.* true or false or open
d. true or false or open

---

22 In Frames 19 and 20 we showed that the open or closed nature of an inequation may depend on the domain of the variable. But it is also possible for it to be independent of the domain. For example, the inequation $n + 2 > n + 1$ is (true/false/open) ________ for all domains of the variable, and the equation $n + 2 = n + 1$ is (true/false/open) ________ for all domains of the variable.

---

true; false

---

23 Label the following sentences first as equations, equalities, inequations, or inequalities and second as true, false, or open:

a. $5 \neq \frac{1}{2}$

b. $3p + 2 = 5$

c. $6k \geqslant \frac{24}{4}$, where k is a positive integer

d. $n + 5 < n + 3$

e. $(5 \times 6) + 1 = 9$

f. $y + 3 = 7$, where y is a negative integer

a. inequality, true *b.* equation, open *c.* inequation, true
d. inequation, false *e.* equality, false *f.* equation, false

24 Label the following sentences first as equations, equalities, inequations, or inequalities and second as true, false, or open:

a. $q > 3$
b. $12 + 5 = 17$
c. $3k \leqslant 6$, where the domain of k is $\{2, 3, 4\}$
d. $7 < -4 + 6$
e. $m - 3 = m - 3$
f. $5q + 1 > -5$, where the domain of q is $\{\ldots -4, -3, -2\}$

a. inequation, open *b.* equality, true *c.* inequation, open
d. inequality, false *e.* equation, true *f.* inequation, false

You have finished the program for Sec. 1-5. Now take the posttest to see how much you have learned.

POSTTEST 1-5

1. Which of the following expressions contains a variable?

   a. $(6 \times k) + (5 \times k)$, where k is -1, 0, or 1
   b. $7p + 9$, where p is a real number
   c. $(4 \times 3) + (5 \times 4)$

2. What are the domains of the variables in Question 1?

3. Label the following sentences as equations, equalities, inequations, or inequalities:

   a. $6q \neq 13$ *b.* $8 + 4 = 15$
   c. $-3 > -5$ *d.* $10 + 4 \neq 15$
   e. $6z \leqslant -3$ *f.* $4y = 12$

4. Which of the sentences in Question 3 are open sentences, and which are closed sentences? Which are true, and which are false?

Which of the following sentences are open, which are true, and which are false?

5. $q + 6 \geqslant q + 1$

6. $k - 3 = k - 2$

7. $7p > 20$, when the domain of p is $\{1, 2, 3\}$

8. $3m \leqslant 9$, when the domain of m is $\{\ldots 1, 2, 3\}$

When you have finished the posttest, correct it yourself using the answers on pages 58 and 59. Then check all the questions you got wrong on the list below:

1. _____ 5. _____

2. _____ 6. _____

3. _____ 7. _____

4. _____ 8. _____

If you compare your performance with that on Pretest 1-5, you will see how much you have improved. If you made more than one mistake on this posttest, review the material and take the pretest again. Don't move on to the next section until you can get all the questions right.

Now proceed to the next section whose number you circled on page 3. If no other sections were circled, take Final Test 1 (page 54).

Section 1-6
Solution Sets and Graphs of Open Sentences

Do this section if you made a mistake in Questions 9 to 13 of Diagnostic Test 1. Begin by taking the pretest below, which will show you how much of Sec. 1-6 you already know and what you will be able to do by the time you have finished it.

PRETEST 1-6

Give the solution sets of the following equations (*a*) when the domain of x is $\{1, 3, 5, 7, 9\}$ and (*b*) when the domain of x is $\{x \mid x \text{ is a real number}\}$:

1. $x + 3 = 10$ *a.* ________ *b.* ________
2. $x + 5 = 9$ *a.* ________ *b.* ________
3. $x + 1 = x + 2$ *a.* ________ *b.* ________
4. $x + 1 = x + 1$ *a.* ________ *b.* ________

If the domain of k is $\{3, 4, 5, 6, 7, 8\}$, draw the graphs of the following sentences or indicate that they have no solutions:

5. $k + 3 < 8$
6. $k + 4 = 6$
7. $k + 5 > 15$
8. $k + 7 = 13$

When you have finished the pretest, correct it yourself using the answers on page 59. Then check all the questions you got wrong on the list below.

1*a.* ____	3*a.* ____	5. ____
1*b.* ____	3*b.* ____	6. ____
2*a.* ____	4*a.* ____	7. ____
2*b.* ____	4*b.* ____	8. ____

If you made any mistakes (even if it was only *one* mistake), read the introduction and go through the program.

If you made no mistakes at all, skip Sec. 1-6, because you already know its material, and go on to take Final Test 1 (page 54).

Introduction

First this section explains the idea of a solution of an open sentence and the concept of a solution set. Then the graph of an open sentence is defined as the graph of its solution set. You will practice with several equations and inequations and learn how to solve and graph them. Special cases are included to show the solution set as the domain of the variable or the empty set and to show how the solution set can alter with the domain of the variable.

FRAMES FOR SECTION 1-6

1 Consider the open sentence $x + 3 > 5$ when the domain of x is $\{1, 2, 3, 4, 5, 6, 7, 8\}$. Since x can represent any element in its domain, this open sentence is equivalent to ______ (how many?) statements. Will all these statements be true? Will all these statements be false?

8; no; no

---

2 The domain of x, $\{1, 2, 3, 4, 5, 6, 7, 8\}$, can be divided into two subsets: those elements of the domain which make the open sentence $x + 3 > 5$ true, that is, {____________}, and those elements of the domain which make the open sentence $x + 3 > 5$ false, that is, {__________}.

$\{3, 4, 5, 6, 7, 8\}$ $\{1, 2\}$

---

3 What is:

a. The union of these two subsets?
b. The intersection of these two subsets?

---

a. $\{1, 2, 3, 4, 5, 6, 7, 8\}$, or the domain of the variable *b.* $\emptyset$

---

4 Those values of the variable which give true statements on substitution in an open sentence are said to *satisfy* the open sentence. Thus 3, 4, 5, 6, 7, and 8 satisfy $x + 3 > 5$, but 1 and 2 do not satisfy $x + 3 > 5$.

a. If the domain of x is $\{2, 4, 6, 8, 10\}$, which values of x satisfy $x + 4 < 10$?
b. Which values of x do not satisfy $x + 4 < 10$?

---

a. 2 and 4 *b.* 6, 8, and 10

---

5 Any value of the variable which satisfies an open sentence is called a *solution* of that open sentence; and the set of all such solutions is called the *solution set* of the open sentence. If the domain of n is $\{0, 1, 2, 3, 4, 5, 6\}$, what are the solutions of $n > 2$? What is the solution set of $n > 2$?

3, 4, 5, and 6 $\{3, 4, 5, 6\}$

---

6 If the domain of k is $\{2, 3, 4, 5, 6\}$, then:

a. What is the solution set of $k + 1 > 5$?
b. What value(s) of k satisfy $k + 1 > 5$?

---

a. $\{5, 6\}$ or $\{k \mid k \text{ is an integer and } 4 < k < 7\}$ *b.* 5 and 6

---

7 If the domain of k is the set of integers, what is the solution set of $k + 1 > 5$?

$\{n \mid n \text{ is an integer and } n > 4\}$ or $\{5, 6, 7, 8 \ldots\}$

---

8 The solution set of an equation or inequation, which is a closed sentence, is defined in the same manner as for an open sentence (Frames 4 and 5); i.e., it is *the set of all values of the variable which give true statements on substitution.* If the domain of n is $\{2, 3, 4, 5\}$, then four statements can be obtained by substitution for n in the inequation $n + 2 > n + 1$.

a. How many of these statements are true?
b. What is the solution set of the inequation?
c. If the domain of n was $\{2, 3, 4, 5, 6, 7, 8\}$, what would the solution set be?

---

a. all four of them *b.* $\{2, 3, 4, 5\}$ *c.* $\{2, 3, 4, 5, 6, 7, 8\}$

---

9 When the domain of n is $\{n \mid n \text{ is a positive integer}\}$, an infinite number of statements can be obtained by substitution for n in the inequation $n + 2 > n + 1$.

a. How many of these statements are true?
b. What is the solution set of the inequation?

---

a. all of them *b.* $\{n \mid n \text{ is a positive integer}\}$

---

10 *a.* If the domain of n is unspecified, i.e., the set of real numbers, what is the solution set of $n + 2 > n + 1$?
b. What is the solution set of $n + 2 = n + 1$?

---

a. $\{n \mid n \text{ is a real number}\}$ *b.* $\emptyset$

---

11 *a.* $n + 2 > n + 1$ is a(n) (true/false/open) __________ sentence, and therefore its solution set is the __________ of the variable.
b. $n + 2 = n + 1$ is a(n) (true/false/open) __________ sentence, and therefore its solution set is __________.

---

a. true, domain *b.* false, $\emptyset$

---

12 *a.* What is the solution set of $y + 3 < y + 2$?
b. What is the solution set of $n + 5 \geqslant n - 1$?

---

a. $\emptyset$ *b.* $\{n \mid n \text{ is a real number}\}$

---

13 What is the solution set of $k + 2 < k + 8$:

a. When the domain of k is $\{k \mid k \text{ is a negative integer}\}$?
b. When the domain of k is $\{-1, 0, 1, 2\}$?

---

a. $\{k \mid k \text{ is a negative integer}\}$ *b.* $\{-1, 0, 1, 2,\}$

---

14 What is the solution set of:

a. $y - 1 < y - 2$
b. $z \leqslant z + 1$

---

a. $\emptyset$ *b.* $\{z \mid z \text{ is a real number}\}$

---

15 A set of numbers such as $\{2, 3, 4\}$ can be graphed on the number line.

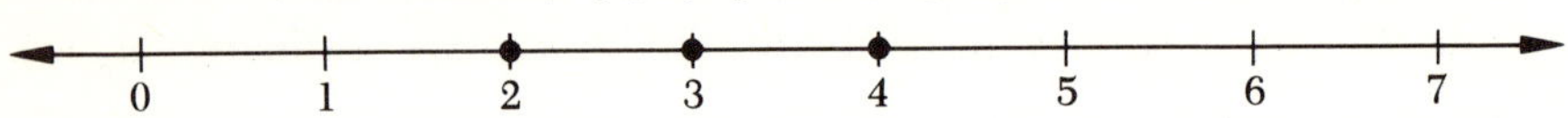

Graph the set $\{-3, -2, -1\}$.

---

16 We define the *graph* of an open sentence as the graph of its solution set. Thus if the domain of k is $\{1, 2, 3, 4, 5, 6\}$, the graph of $k + 3 > 6$ is

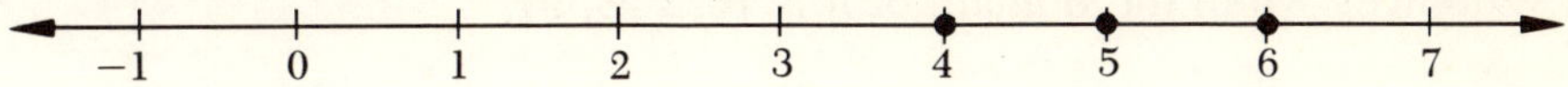

Draw the graph of $k + 3 > 6$ when the domain of k is $\{1, 2, 3, 4, 5, 6, 7\}$.

17 Draw the graph of $k + 2 < 5$ when the domain of k is $\{1, 2, 3, 4, 5\}$. **Hint**: See whether each value of k gives a true or false statement.

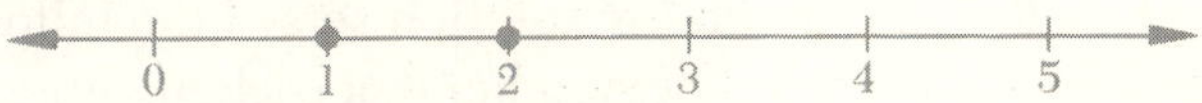

18 Draw the graph of $k + 2 = 5$ when the domain of k is $\{1, 2, 3, 4, 5\}$.

19 When the domain of k is $\{-3, -2, -1, 0, 1, 2\}$, draw the graph of $k + 2 = 1$.

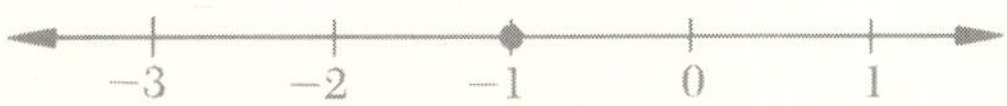

20 When the domain of k is $\{-3, -2, -1, 0, 1, 2\}$, draw the graph of $k + 2 < 1$.

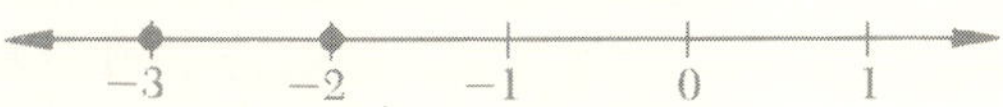

21 If the domain of k is $\{-3, -2, -1, 0, 1, 2, 3\}$, draw the graphs of

a. $k + 3 = 1$
b. $k + 2 < 1$
c. $k + 3 > -3$

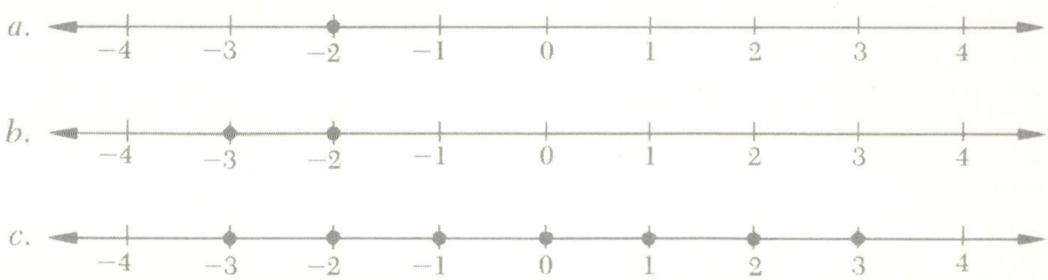

22 Each solution must be in the domain of the variable because the solution set is defined as a subset of the domain. Give the solution sets of the following sentences when the domain of n is $\{6, 7, 8, 9\}$:

a. $n + 1 > 8$
b. $n + 5 = 10$
c. $n + 2 \leqslant 7$
d. $n + 4 = 13$

---

a. $\{8, 9\}$ *b.* $\emptyset$ *c.* $\emptyset$ *d.* $\{9\}$

---

23 If the domain of q is $\{1, 3, 5, 7, 9\}$, graph the solution sets of the follow- sentences. Write $\emptyset$ if there are no solutions. (Remember that each member of the solution set *must* be in the domain of the variable.)

a. $q + 2 = 6$
b. $q + 2 = 9$
c. $q + 6 = q + 3 + 3$
d. $q + 2 < 2$

---

a. $\emptyset$

b.
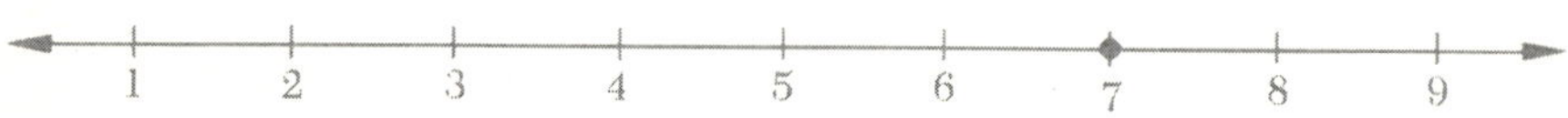

c.

d. $\emptyset$

---

24 If the domain of y is $\{0, 2, 4, 6, 8\}$, graph the solution sets of the following sentences. Write $\emptyset$ if there are no solutions:

a. $y + 1 = 5$
b. $y + 2 = 5$
c. $y + 4 < 2$
d. $y + 5 + 2 = y + 6 + 1$

---

a.
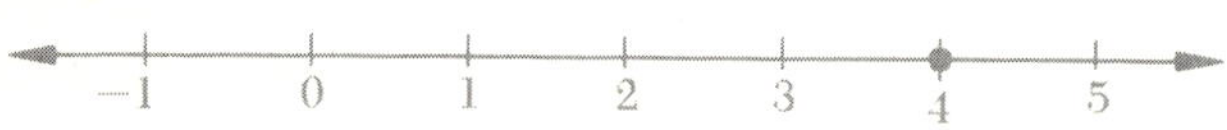

b. $\emptyset$ *c.* $\emptyset$

d.

---

25 If y is a variable over the set of real numbers, find the solution sets of:

a. $y + 4 = y + 5$
b. $y + 4 + 1 = y + 5$

---

a. $\emptyset$ *b.* $\{y \mid y \text{ is a real number}\}$

---

26 Find the solution set of $y + 7 = 12$ when the domain of y is

a. $\{1, 2, 3, 4, 5\}$
b. $\{y \mid y \text{ is a rational number}\}$

---

a. $\{5\}$ *b.* $\{5\}$

---

27 Find the solution set of $y + 6 = 12$ when the domain of y is:

a. $\{1, 2, 3, 4, 5\}$
b. $\{y \mid y \text{ is a rational number}\}$

---

a. $\emptyset$ *b.* $\{6\}$

---

28 Give the solution set of $k + 5 = k + 4 + 1$ when the domain of k is:

a. $\{1, 2, 3, 4, 5\}$
b. $\{k \mid k \text{ is a rational number}\}$

---

a. $\{1, 2, 3, 4, 5\}$ *b.* $\{k \mid k \text{ is a rational number}\}$

You have finished the program for Sec. 1-6. Now take the posttest to see how much you have learned.

POSTTEST 1-6

Give the solution sets of the following equations (*a*) when the domain of x is $\{2, 4, 6, 8, 10\}$ and (*b*) when the domain of x is $\{x \mid x \text{ is a real number}\}$:

1. $x + 8 = 14$ *a.* ________ *b.* ________
2. $x + 6 = 11$ *a.* ________ *b.* ________
3. $x + 3 = x + 4$ *a.* ________ *b.* ________
4. $x + 3 = x + 3$ *a.* ________ *b.* ________

If the domain of k is $\{3, 4, 5, 6, 7, 8\}$, draw the graph of the following sentences or indicate that they have no solutions:

5. $k + 2 < 5$
6. $k + 4 = 8$
7. $k + 7 > 13$
8. $k + 6 = 15$

When you have finished the posttest, correct it yourself using the answers on page 59. Then check all the questions you got wrong on the list below:

1*a*. ______	3*a*. ______	5. ______
1*b*. ______	3*b*. ______	6. ______
2*a*. ______	4*a*. ______	7. ______
2*b*. ______	4*b*. ______	8. ______

If you compare your performance with that on Pretest 1-6, you will see how much you have improved. If you made more than one mistake on this posttest, review the material and take the pretest again. Don't move on until you can get all the questions right.

Now proceed to Final Test 1.

FINAL TEST 1

1. Which of the following statements are true?

   a. $-14 \neq -7 \times 2$
   b. $-3 < -5$
   c. $-2 \leqslant -8 \div 4$
   d. $8 \geqslant -3 \times 2$

2. If n is an integer and $-2 < n \leqslant 3$, what values can n have?

3. Which of the numbers $\sqrt{50}$, $\frac{-5}{6}$, $-\sqrt{9}$, $\frac{0}{3}$, $\frac{-4}{2}$, -3 are:

   a. real numbers
   b. rational numbers
   c. irrational numbers
   d. integers

4. Write the following sets in element notation:

   a. $\{m \mid m \text{ is an integer and } m > -2\}$
   b. $(-5, -4, -3, -2\} \cup \{-3, -2, 2, 3\}$
   c. $\{1, 2, 3, 4\} \cap \{4, 5, 6, 7\}$

5. Write the following sets in property notation:

   a. $\{-15, -14, -13, -12, -11\}$
   b. $\{y \mid y \text{ is a rational number}\} \cap \{z \mid z \text{ is an irrational number}\}$

c. $\{q \mid q \text{ is a positive rational number}\} \cup \{p \mid p \text{ is a rational number}\}$
d. $\{m \mid m \text{ is an integer}\} \cap \{n \mid n \text{ is a rational number}\}$

6. Which of the following sentences are closed, and which are open?

a. $y + 5 = 9$	*b.* $m + 5 = m + 5$
c. $(2 \times 3) + 1 = 9$	*d.* $10 > x + 2$
e. $w - 3 < w - 8$	*f.* $-2 \times (-3) = 6$

7. Which of the sentences in Question 6 are true?

Suppose x is a variable over the set of numbers $\{0, 2, 4, 6, 8, 10\}$:

8. What is the domain of x?

9. Graph the solution set of the open sentence $x + 3 = 11$

10. Graph the solution set of the open sentence $x > 6$

Give the solution sets of the following equations (*a*) when the domain of x is $\{1, 3, 5, 7, 9\}$ and (*b*) when the domain of x is $\{x \mid x \text{ is a real number}\}$:

11. $x + 6 = 8$ *a.* __________ *b.* __________

12. $x + 5 = x + 6$ *a.* __________ *b.* __________

13. $x + 5 = x + 5$ *a.* __________ *b.* __________

The answers for this test are on page 60. Check the questions you got wrong on the list below and compare your results with those on Diagnostic Test 1 to see how much you have improved.

1. _____	4*a*. _____	5*d*. _____	11*a*. _____
2. _____	4*b*. _____	6. _____	11*b*. _____
3*a*. _____	4*c*. _____	7. _____	12*a*. _____
3*b*. _____	5*a*. _____	8. _____	12*b*. _____
3*c*. _____	5*b*. _____	9. _____	13*a*. _____
3*d*. _____	5*c*. _____	10. _____	13*b*. _____

If you made more than one mistake, review the material and take the diagnostic test again. Make sure that you can get all the questions right before you finish the chapter.

Now proceed to Chap. 2.

ANSWERS TO TESTS FOR CHAPTER 1

DIAGNOSTIC TEST 1

1. c and d

2. $-3, -2, -1, 0, 1$

3. *a.* $\sqrt{30}$ $\quad \frac{7}{-3}$ $\quad -\sqrt{16}$ $\quad \frac{0}{5}$ $\quad \frac{-6}{3}$ $\quad -4$

   b. $\frac{7}{-3}$ $\quad \frac{0}{5}$ $\quad \frac{-6}{3}$ $\quad -\sqrt{16}$ $\quad -4$

   c. $\sqrt{30}$

   d. $-\sqrt{16}$ $\quad \frac{-6}{3}$ $\quad \frac{0}{5}$ $\quad -4$

4. *a.* $\{\ldots, -1, 0, 1, 2, 3\}$
   b. $\{-2, -1, 0, 1, 2, 3, 4\}$
   c. $\{5\}$

5. *a.* $\{n \mid n$ is an integer and $-18 < n < -12\}$
   b. $\emptyset$
   c. $\{y \mid y$ is a real number$\}$
   d. $\{p \mid p$ is a negative integer$\}$

6. *a.* open
   b. closed
   c. closed
   d. open
   e. closed
   f. closed

7. e and f

8. $\{1, 3, 5, 7, 9, 11\}$

9. 

10. 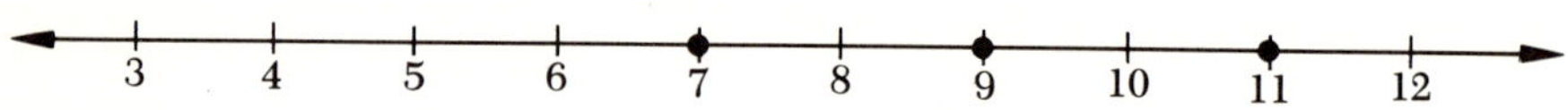

11. *a.* $\emptyset$
    b. 3

12. *a.* $\emptyset$
    b. $\emptyset$

13. *a.* $\{0, 2, 4, 6, 8, 10\}$
    b. $\{x \mid x$ is a real number$\}$

PRETEST 1-1

1. false
2. false
3. false
4. true
5. true
6. true
7. yes
8. yes
9. $\{-4, -3, -2, -1, 0\}$
10. $-11 \leqslant n < -2$

POSTTEST 1-1

1. true
2. true
3. true
4. false
5. false
6. false
7. yes
8. yes
9. $\{0, 1, 2, 3, 4\}$
10. $-8 < n \leqslant -3$

PRETEST 1-2

1. Real numbers: $-\sqrt{5}$ $\sqrt{36}$ $\frac{0}{-3}$ $\frac{-8}{2}$ $\frac{10}{-12}$ -6
2. Rational numbers: $\frac{10}{-12}$ $\sqrt{36}$ $\frac{0}{-3}$ $\frac{-8}{2}$ -6
3. Irrational numbers: $-\sqrt{5}$
4. Integers: $\sqrt{36}$ $\frac{0}{-3}$ $\frac{-8}{2}$ -6

POSTTEST 1-2

1. Real numbers: $\sqrt{25}$ $-\sqrt{6}$ $\frac{0}{-5}$ $\frac{-10}{2}$ $\frac{12}{-36}$ -8
2. Rational numbers: $\frac{12}{-36}$ $\sqrt{25}$ $\frac{0}{-5}$ $\frac{-10}{2}$ -8
3. Irrational numbers: $-\sqrt{6}$
4. Integers: $\sqrt{25}$ $\frac{0}{-5}$ $\frac{-10}{2}$ -8

PRETEST 1-3

1. *a* and *b*
2. $\{\ldots, -3, -2, -1\}$
3. $\{\ldots, -1, 0, 1, 2\}$
4. $\{n \mid n$ is an integer and $n > 4\}$
5. $\{n \mid n$ is an integer and $-4 < n < 4\}$
6. 2, 3, 4

POSTTEST 1-3

1. *a* and *b*
2. $\{1, 2, 3, \ldots\}$
3. $\{-1, 0, 1, 2, \ldots\}$
4. $\{n \mid n$ is an integer and $n > 2\}$
5. $\{n \mid n$ is an integer and $-6 < n < 2\}$
6. 5

PRETEST 1-4

1. no
2. no
3. $\{-5, -4, 1, 2, 3, 4, 5\}$
4. $\{-1, 0\}$
5. $\emptyset$
6. $\{c \mid c$ is a rational number$\}$
7. $\{e \mid e$ is an irrational number$\}$

POSTTEST 1-4

1. no
2. yes
3. $\{4\}$
4. $\{-4, -3, -2, -1, 0, 1, 2, 3\}$
5. $\emptyset$
6. $\{d \mid d$ is an integer$\}$
7. $\{e \mid e$ is a rational number$\}$

PRETEST 1-5

1. *a* and *c*
2. *a.* set of rational numbers
   c. $\{1, 2, 3\}$
3. *a.* inequality
   b. inequation
   c. equation
   d. inequation
   e. equality
   f. inequality

POSTTEST 1-5

1. *a* and *b*
2. *a.* $\{-1, 0, 1\}$
   b. set of real numbers
3. *a.* inequation
   b. equality
   c. inequality
   d. inequality
   e. inequation
   f. equation

4. *a*, *e*, *f* are closed sentences
   b, *c*, *d* are open sentences
   a, *f* are true sentences
   e is a false sentence

5. true

6. false

7. open

8. true

PRETEST 1-6

1. *a.* {7}
   b. {7}

2. *a.* ∅
   b. {4}

3. *a.* ∅
   b. ∅

4. *a.* {1, 3, 5, 7, 9}
   b. {*x* | *x* is a real number}

5. 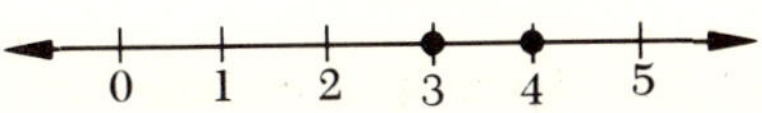

6. ∅

7. ∅

8. 

4. *a*, *e*, *f* are open sentences
   b, *c*, *d* are closed sentences
   c, *d* are true sentences
   b is a false sentence

5. true

6. false

7. open

8. true

POSTTEST 1-6

1. *a.* {6}
   b. {6}

2. *a.* ∅
   b. 5

3. *a.* ∅
   b. ∅

4. *a.* {2, 4, 6, 8, 10}
   b. {*x* | *x* is a real number}

5. ∅

6. 

7. 

8. ∅

FINAL TEST 1

1. c and d

2. $-1, 0, 1, 2,$ and 3

3. *a.* all of them

   b. $\frac{-5}{6}$ $\frac{0}{3}$ $\frac{-4}{2}$ $-\sqrt{9}$ -3

   c. $\sqrt{50}$

   d. $\frac{0}{3}$ $\frac{-4}{2}$ $-\sqrt{9}$ -3

4. *a.* $\{-1, 0, 1, 2, 3, \ldots\}$
   b. $\{-5, -4, -3, -2, 2, 3\}$
   c. $\{4\}$

5. *a.* $\{n \mid n \text{ is an integer and } -16 < n < -10\}$
   b. $\emptyset$
   c. $\{p \mid p \text{ is a rational number}\}$
   d. $\{m \mid m \text{ is an integer}\}$

6. *a.* open
   b. closed
   c. closed
   d. open
   e. closed
   f. closed

7. b and f

8. $\{0, 2, 4, 6, 8, 10\}$

9. 

10. 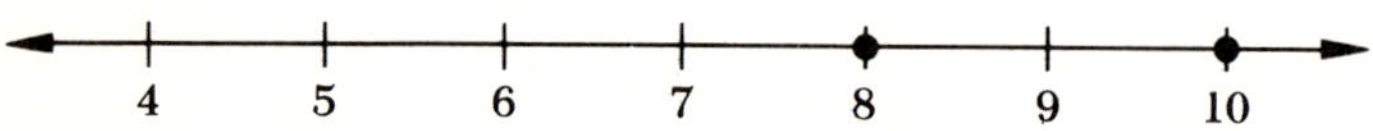

11. *a.* $\emptyset$
    b. $\{2\}$

12. *a.* $\emptyset$
    b. $\emptyset$

13. *a.* $\{1, 3, 5, 7, 9\}$
    b. $\{x \mid x \text{ is a real number}\}$

Chapter 2
First-degree Equations in One Variable

In this chapter you will learn how to solve first-degree equations in one variable, including all the special cases. Section 2-1 introduces equivalent equations, i.e., equations having the same solution set. It then shows that adding the same real number to both sides of an equation gives an equivalent equation. Section 2-2 shows that multiplication of both sides of an equation by the same real number gives an equivalent equation and Sec. 2-3 that addition of the same multiple of the variable to both sides of an equation gives an equivalent equation. Section 2-4 extends these methods to the solution of equations with more than three terms. Section 2-5 discusses equations containing parentheses but not fractions, Sec. 2-6 those containing fractions but not parentheses, and Sec. 2-7 those containing both. Finally Sec. 2-8 covers the special cases which give solutions sets of $\{0\}$, $\emptyset$, or the domain of the variable.

DIAGNOSTIC TEST 2

Give the solution sets of the following equations and check your answers.

1. $\frac{2}{3}k = -\frac{5}{7}$

2. $-\frac{8y}{5} = -\frac{10}{9}$

3. $4y + 5 - 8y = 10y - 6 + 9$

4. $7b - 6 + 2 = 8b + 7 - 1 - 11b$

5. $4(3y - 2) - 6(7 - 5y) = 12y - 25$

6. $6(4n - 3) - (n + 2) = 3(4 - 5n)$

7. $\frac{5y}{7} - \frac{2}{3} + \frac{y}{6} = \frac{9y}{14} + \frac{2}{7}$

8. $\frac{1}{3}(5y + 2) - \frac{2}{5}(3y - 4) = \frac{1}{15}(4y + 25)$

9. $\frac{1}{2}\left(y + \frac{2}{3}\right) - \frac{4}{3}\left(\frac{3y}{4} + 5\right) = -\frac{1}{2}(35y + 24)$

10. $3(5y - 2) + 4(2 - 7y) = 2(1 - 3y)$

11. $3y + 2(1 - 2y) = 10 - 8 - y$

12. $\frac{2}{3}(3y + 5) - \frac{3}{7}(5y - 2) = -\frac{1}{7}(y + 6)$

When you have finished the diagnostic test, correct it yourself, using the answers on page 133. Then check all the questions you got wrong on the list below:

1. ______	5. ______	7. ______	10. ______
2. ______	6. ______	8. ______	11. ______
3. ______		9. ______	12. ______
4. ______			

If you made no mistakes at all, you need not do any work in this chapter and should go straight on to Chap. 3, First-degree Inequations in One Variable, which begins on page 137. But if you made any mistakes (even if it was only *one* mistake) follow these instructions:

	section	page
If you made any mistakes in Questions 1 to 4, circle Secs. 2-1 to 2-4.	2-1	63
	2-2	70
	2-3	80
	2-4	90
If you made any mistakes in Question 5 or 6, circle Sec. 2-5.	2-5	100
If you made any mistakes in Questions 7 to 9, circle Secs. 2-2, 2-6, and 2-7.	2-6	107
	2-7	116
If you made any mistakes in Questions 10 to 12, circle Sec. 2-8.	2-8	126

You should now work through the sections whose numbers you have circled, taking them in numerical order. The sections begin on the pages indicated.

Section 2-1
The Addition Transformation

Do this section if you made a mistake in Questions 1 to 4 of Diagnostic Test 2. Begin by taking the pretest below, which will show you how much of Sec. 2-1 you already know and what you will be able to do by the time you have finished it.

PRETEST 2-1

Convert the following equations to equivalent equations of the form $z = ?$:

1. $z + 6 = -10$
2. $z - 2 = 7$
3. $z + 4 = 6$

Give the solution sets of the following equations:

4. $k - 7 = -3$
5. $y - 6 = 13$
6. $q + 5 = -9$

When you have finished the pretest, correct it yourself using the answers on page 133. Then check all the questions you got wrong on the list below:

1. ______ 4. ______

2. ______ 5. ______

3. ______ 6. ______

If you made any mistakes (even if it was only *one* mistake), read the introduction and go through the program.

If you made no mistakes at all, skip Sec. 2-1, because you already know its material, and go on to Sec. 2-2.

Introduction

Section 2-1 begins by showing you that an open sentence such as $y + 5 = 7$ yields a separate statement for each value of variable. Values of the variable which yield true statements are called *solutions.* You learn that open sentences having the same solution set are defined as *equivalent.* You are shown that adding the same real number to both sides of an equation gives an equivalent equation, and you can use this property to solve equations of the general type $x + a = b$.

FRAMES FOR SECTION 2-1

1 *a.* If the domain of y is $\{2, 4, 6, 8\}$, then any open sentence containing y will represent ______ (how many?) different statements.

b. List all the possible statements obtained from the open sentence $y + 2 = 6$.

c. Which of these statements are true?

d. What is the solution set of $y + 2 = 6$?

---

a. four *b.* $2 + 2 = 6$; $4 + 2 = 6$; $6 + 2 = 6$; $8 + 2 = 6$
c. only $4 + 2 = 6$ *d.* $\{4\}$

---

2 *a.* If the domain of y is $\{2, 4, 6, 8,\}$, list all the possible statements obtained from the open sentence $y + 3 = 7$.

b. Which of these statements are true?

c. What is the solution set of $y + 3 = 7$?

---

a. $2 + 3 = 7$; $4 + 3 = 7$; $6 + 3 = 7$; $8 + 3 = 7$ *b.* only $4 + 3 = 7$
c. $\{4\}$

---

3 *a.* If the domain of y is $\{2, 4, 6, 8\}$, list all the possible statements obtained from the open sentence $y = 4$.

b. Which of these statements are true?

c. What is the solution set of $y = 4$?

---

a. $2 = 4$; $4 = 4$; $6 = 4$; $8 = 4$ *b.* only $4 = 4$ *c.* $\{4\}$

---

4 If the domain of y is $\{2, 4, 6, 8\}$, what are the solution sets of the following open sentences?

a. $y = 4$
b. $y + 1 = 5$
c. $y + 2 = 6$
d. $y + 3 = 7$
e. $y + 4 = 8$

---

a. $\{4\}$ *b.* $\{4\}$ *c.* $\{4\}$ *d.* $\{4\}$ *e.* $\{4\}$

---

5 If two open sentences have the same solution set, they are said to be *equivalent*. If the domain of k is $\{1, 2, 3, 4\}$, which of the following pairs of open sentences are equivalent?

a. $k = 1$ and $k = 2$
b. $k = 1$ and $k + 1 = 2$
c. $k = 1$ and $k + 1 = 3$
d. $k = 2$ and $k + 1 = 3$
e. $k = 1$ and $k + 2 = 3$

---

b, *d*, and *e*

---

6 The sign $\longleftrightarrow$ is placed between two open sentences to show that they are equivalent. Which of the following statements are true if the domain of p is $\{0, 1, 2, 3\}$?

a. $p = 2 \longleftrightarrow p = 3$
b. $p = 2 \longleftrightarrow p + 2 = 4$
c. $p = 2 \longleftrightarrow p + 2 = 5$
d. $p = 2 \longleftrightarrow p + 2 = 6$
e. $p = 3 \longleftrightarrow p + 2 = 4$
f. $p = 3 \longleftrightarrow p + 2 = 5$
g. $p = 3 \longleftrightarrow p + 2 = 6$

---

b and *f*

---

7 *a.* Write the true statement which can be obtained from the open sentence $z = 5$ if the domain of z is the set of real numbers.
b. What is the solution set of $z = 5$?

---

a. $5 = 5$ *b.* $\{5\}$

---

8 *a.* Write the true statement which can be obtained from the open sentence $z + 1 = 6$ if the domain of z is the set of real numbers.
b. What is the solution set of $z + 1 = 6$?

---

a. $5 + 1 = 6$ *b.* $\{5\}$

---

9 If the domain of n is the set of real numbers, what are the solution sets of the following open sentences?

a. $n = 6$
b. $n + 1 = 7$
c. $n + 2 = 8$

---

a. $\{6\}$ *b.* $\{6\}$ *c.* $\{6\}$

---

10 If the domain of k is the set of real numbers,

$k = 7 \longleftrightarrow k + \_ = 8 \longleftrightarrow k + 2 = \_$

---

1 9

---

11 The answer to the previous frame can be written in the form

$k = 7 \longleftrightarrow k + 1 = 7 + 1 \longleftrightarrow k + 2 = 7 + 2$

Similarly it can be shown that:

a. $k = 7 \longleftrightarrow k + 11 = 7 + \_\_\_\_$

b. $k = 7 \longleftrightarrow k + 39 = 7 + \_\_\_\_$

c. $k = 7 \longleftrightarrow k + m = 7 + \_\_\_\_$, where m is any real number

---

a. 11 *b.* 39 *c.* m

---

12 If you take an open sentence such as $k = 8$ and add the same real number n to the expressions on each side of the equals sign, you obtain a second open sentence $k + n = 8 + n$. What can you say about these two open sentences?

They are equivalent, or they have the same solution set.

---

13 If you take an open sentence such as $k + 4 = 10$ and add -4 to the expressions on each side of the equals sign, you obtain a second open sentence

$k + 4 + (-4) = 10 + (-4)$

or $k = \_\_\_\_$

What can you say about these two open sentences?

---

6 They are equivalent, or they have the same solution set.

---

14 If you take an open sentence such as $k - 5 = 11$ and add 5 to the expressions on each side of the equals sign, you obtain a second open sentence $k =$ ______.
What can you say about these two open sentences?

---

16 They are equivalent, or they have the same solution set.

---

15 In general, whenever you add the same number to each side of an equation, you obtain a(n) ______________ equation.

---

equivalent

---

16 Adding a number to each side can be a good method of solving an equation, i.e., finding its solution.

a. $x - 2 = 7 \longleftrightarrow x - 2 + 2 = 7 + 2 \longleftrightarrow x =$ ___ adding 2 to each side

Solution set is ______.

b. $y + 5 = -2 \longleftrightarrow y + 5 +$ ___ $= -2 +$ ___ $\longleftrightarrow y =$ ___ adding -5 to each side

Solution set is ______.

c. $d - 4 = -4 \longleftrightarrow d - 4 +$ ___ $= -4 +$ ___ $\longleftrightarrow d =$ ___ adding 4 to each side

Solution set is ______.

---

a. 9; {9} *b.* −5; −5; −7; {−7} *c.* 4; 4; 0; {0}

---

17 *a.* $k + 3 = -2 \longleftrightarrow k =$ ______ adding -3 to each side
Solution set is ______.

b. $x - 5 = -3 \longleftrightarrow x =$ ______ adding ______ to each side
Solution set is ______.

c. $b + 7 = -11 \longleftrightarrow b =$ ______ adding ______ to each side
Solution set is ______.

---

a. −5; {−5} *b.* 2; 5; {2} *c.* −18; −7; {−18}

---

18 Use the method of adding a number to each side to solve the following equations:

a. $q + 6 = -1 \longleftrightarrow q =$ _____ adding _____ to each side
Solution set is _____.

b. $m - 7 = 6 \longleftrightarrow m =$ _____
Solution set is _____.

c. $x + 12 = -5 \longleftrightarrow x =$ _____
Solution set is _____.

---

a. -7; -6; $\{-7\}$ *b.* 13; $\{13\}$ *c.* -17; $\{-17\}$

---

19 Solve the following equations:

a. $x + 13 = -4 \longleftrightarrow x =$ _____
Solution set is _____.

b. $y - 11 = -8 \longleftrightarrow y =$ _____
Solution set is _____.

---

a. -17; $\{-17\}$ *b.* 3; $\{3\}$

---

20 Solve the following equations:

a. $m - 10 = -20$
b. $p + 17 = -4$

---

a. $\{-10\}$ *b.* $\{-21\}$

You have finished the program for Sec. 2-1. Now take the posttest to see how much you have learned.

POSTTEST 2-1

Convert the following equations to equivalent equations of the form $z = ?$:

1. $z + 2 = -7$
2. $z - 5 = 6$
3. $z + 5 = 9$

Give the solution sets of the following equations:

4. $k - 3 = -6$
5. $y - 2 = 11$
6. $q + 6 = -4$

When you have finished the posttest, correct it yourself using the answers on page 133. Then check all the questions you got wrong on the list below:

1. ______	4. ______
2. ______	5. ______
3. ______	6. ______

If you compare your performance with that on Pretest 2-1, you will see how much you have improved. If you made more than one mistake on this posttest, review the material and take the pretest again. Don't move on to the next section until you get all the questions right.

Now proceed to Sec. 2-2.

Section 2-2
The Multiplication Transformation

Do this section if you made a mistake in Questions 1 to 4 or 7 to 9 of Diagnostic Test 2. Begin by taking the pretest below, which will show you how much of Sec. 2-2 you already know and what you will be able to do by the time you have finished it.

PRETEST 2-2

Convert the following equations to equivalent equations of the form $k = ?$:

1. $6k = -3 \longleftrightarrow k =$ ______

2. $-\frac{2}{7}k = \frac{2}{3} \longleftrightarrow k =$ ______

3. $-15k = -5 \longleftrightarrow k =$ ______

Give the solution sets of the following equations:

4. $8y = \frac{13}{2}$

5. $-\frac{3}{4}z = -5$

6. $\frac{2}{3}q = 7$

When you have finished the pretest, correct it yourself using the answers on page 134. Then check all the questions you got wrong on the list below:

1. ______	3. ______	5. ______
2. ______	4. ______	6. ______

If you made any mistakes (even if it was only *one* mistake), read the introduction and go through the program.

If you made no mistakes at all, skip Sec. 2-2, because you already know its material, and go on to the next section whose number you circled on page 62.

Introduction

First you learn the notation $3z$ for $3 \times z$, $-5f$ for $-5 \times f$, $-\frac{2}{3}y$ for $-\frac{2}{3} \times y$, etc. Then you are shown that multiplying both sides of an equation by the same real number yields an equivalent equation. You use this property to solve equations of the general type $\frac{a}{b}y = \frac{c}{d}$.

FRAMES FOR SECTION 2-2

1 $7 \times k$ is written $7k$, $\frac{5}{6} \times q$ as $\frac{5}{6}q$, etc.

How would you write the following?

a. $3 \times q$

b. $\frac{1}{6} \times y$

c. $\frac{7}{5} \times z$

---

a. $3q$ *b.* $\frac{1}{6}y$ *c.* $\frac{7}{5}z$

---

2 Since the operations *negative of* and *multiplying by* -1 are equivalent,

$-3q$ (the negative of $3q$) $= 3q \times (-1) = -1 \times 3q$

Hence

$-3q = -1 \times 3 \times q =$ ______ $\times q$

How would you write the following in the shortest form?

a. $-5 \times k$

b. $-\frac{1}{8} \times z$

c. $-\frac{7}{9} \times w$

-3 *a.* $-5k$ *b.* $-\frac{1}{8}z$ *c.* $-\frac{7}{9}w$

---

3 Since $3 \times p = p \times 3$, they are both written as $3p$. How would you write the following?

a. $q \times 5$

b. $k \times (-8)$

c. $-\frac{1}{3} \times m$

d. $z \times (-2)$

a. $5q$ *b.* $-8k$ *c.* $-\frac{1}{3}m$ *d.* $-2z$

4 *a.* If the domain of y is $\{1, 3, 5, 7, 9\}$ then any open sentence containing y will represent __________ (how many?) different statements.

b. List all the possible statements obtained from the open sentence $6y = 42$.

c. Which of these statements are true?

d. What is the solution set of $6y = 42$?

a. five

b. $6 \times 1 = 42$; $6 \times 3 = 42$; $6 \times 5 = 42$; $6 \times 7 = 42$; $6 \times 9 = 42$

c. only $6 \times 7 = 42$ *d.* $\{7\}$

5 *a.* If the domain of y is $\{1, 3, 5, 7, 9\}$, list all the possible statements obtained from the open sentence $2y = 14$.

b. Which of these statements are true?

c. What is the solution set of $2y = 14$?

a. $2 \times 1 = 14$; $2 \times 3 = 14$; $2 \times 5 = 14$; $2 \times 7 = 14$; $2 \times 9 = 14$

b. only $2 \times 7 = 14$ *c.* $\{7\}$

6 *a.* If the domain of y is $\{1, 3, 5, 7, 9\}$, list all the possible statements obtained from the open sentence $y = 7$.

b. Which of these statements are true?

c. What is the solution set of $y = 7$?

a. $1 = 7$; $3 = 7$; $5 = 7$; $7 = 7$; $9 = 7$ *b.* only $7 = 7$ *c.* $\{7\}$

7 If the domain of y is $\{1, 3, 5, 7, 9\}$, what are the solution sets of the following open sentences?

a. $y = 7$
b. $2y = 14$
c. $3y = 21$
d. $4y = 28$
e. $5y = 35$
f. $6y = 42$

a. $\{7\}$ *b.* $\{7\}$ *c.* $\{7\}$ *d.* $\{7\}$ *e.* $\{7\}$ *f.* $\{7\}$

8 Which of the following statements are true if the domain of y is $\{1, 3, 5, 7, 9\}$?

a. $y = 7 \longleftrightarrow 7y = 49$
b. $y = 3 \longleftrightarrow -3y = -9$
c. $y = 9 \longleftrightarrow \frac{1}{3}y = 3$
d. $y = 5 \longleftrightarrow \frac{1}{5}y = 1$
e. $y = 1 \longleftrightarrow 2y = 1$

---

a, *b*, *c*, and *d* are true; *e* is false

---

9 *a.* Write the true statement which can be obtained from the open sentence $3z = 18$ if the domain of z is the set of real numbers.
b. What is the solution set of $3z = 18$?

---

a. $3 \times 6 = 18$ *b.* $\{6\}$

---

10 If the domain of k is the set of real numbers, what are the solution sets of these open sentences?

a. $k = 9$
b. $5k = 25$
c. $3k = -12$
d. $\frac{1}{3}k = 4$
e. $-2k = 10$

---

a. $\{9\}$ *b.* $\{5\}$ *c.* $\{-4\}$ *d.* $\{12\}$ *e.* $\{-5\}$

---

11 If the domain of k is the set of real numbers,

$k = 3 \longleftrightarrow$ _____$k = 21 \longleftrightarrow 11k =$ _____

---

7 33

---

12 The answer to the previous frame can be written in the form

$$k = 3 \longleftrightarrow 7k = 3 \times 7 \longleftrightarrow 11k = 3 \times 11$$

Similarly it can be shown that:

a. $k = 3 \longleftrightarrow 5k =$ ______ $\times 3$

b. $k = 3 \longleftrightarrow 4k =$ ______ $\times 3$

c. $k = 3 \longleftrightarrow -2k =$ ______ $\times 3$

d. $k = 3 \longleftrightarrow m \times k =$ ______ $\times 3$, where m is any real number and $m \neq 0$

---

a. 5 *b.* 4 *c.* −2 *d.* m

---

13 If you take an open sentence such as $q = 4$ and multiply the expressions on each side of the equals sign by the same real number n ($n \neq 0$), you obtain a second open sentence, $q \times n = 4 \times n$. What can you say about these two open sentences?

They are equivalent, or they have the same solution set.

---

14 *a.* If the domain of q is $\{1, 2, 3, 4\}$, what four statements are obtainable from the open sentence $q \times 0 = 4 \times 0$?

b. Which of these statements are true?

---

a. $1 \times 0 = 4 \times 0$; $2 \times 0 = 4 \times 0$; $3 \times 0 = 4 \times 0$; $4 \times 0 = 4 \times 0$

b. all of them

---

15 If the domain of q is $\{1, 2, 3, 4\}$, what is the solution set of $q \times 0 = 4 \times 0$?

$\{1, 2, 3, 4\}$

---

16 If the domain of k is $\{2, 4, 6, 8\}$, what is the solution set of the open sentence $k \times m = 4m$ when:

a. $m = 0$ (think carefully which values of k give true statements)

b. $m = 1$

c. $m = 2$

d. m is any real number other than 0

---

a. $\{2, 4, 6, 8\}$ *b.* $\{4\}$ *c.* $\{4\}$ *d.* $\{4\}$

---

17 *a.* $q = 5 \longleftrightarrow q \times n = 5 \times n$ is true for all values of n except ______.

b. $q = -2 \longleftrightarrow q \times a = -2a$ is true for all values of a except ______.

---

a. 0 *b.* 0

---

18 If you take an open sentence such as $5k = 30$ and multiply each side of the equation by $\frac{1}{5}$, you obtain a second open sentence $5k \times \frac{1}{5} = 30 \times \frac{1}{5}$ or $k =$ ______.
What can you say about these two open sentences?

---

6 They are equivalent, or they have the same solution set.

---

19 If you take an open sentence such as $\frac{1}{7}k = 6$ and multiply each side of the equation by ______, you obtain a second open sentence $k =$ ______.
What can you say about these two open sentences?

7; 42 They are equivalent, or they have the same solution set.

---

20 In general, whenever you multiply each side of an open sentence by the same number, you obtain a(n) ______ open sentence, provided that the number is not ______.

---

equivalent; zero

---

21 Solve the following equations by using multiplication:

a. $4z = 12 \longleftrightarrow 4z \times \frac{1}{4} = 12 \times \frac{1}{4} \longleftrightarrow z =$ ______

Multiply each side by the reciprocal of 4, which is $\frac{1}{4}$.

Solution set is ______.

b. $\frac{1}{3}y = 7 \longleftrightarrow \frac{1}{3}y \times$ ______ $= 7 \times$ ______ $\longleftrightarrow y =$ ______

Multiply each side by the reciprocal of $\frac{1}{3}$, which is 3.

Solution set is ______.

c. $10d = 5 \longleftrightarrow 10d \times$ ______ $= 5 \times$ ______ $\longleftrightarrow d =$ ______

Multiply each side by the reciprocal of 10, which is ______.

Solution set is ______.

a. 3; {3} *b.* 3; 3; 21; {21} *c.* $\frac{1}{10}$; $\frac{1}{10}$; $\frac{1}{2}$; $\frac{1}{10}$; $\left\{\frac{1}{2}\right\}$

22 Solve the following equations using multiplication:

a. $-5y = 20 \longleftrightarrow -5y \times$ ______ $= 20 \times$ ______ $\longleftrightarrow y =$ ______

Multiply each side by the reciprocal of -5, which is $-\frac{1}{5}$.
Solution set is ______.

b. $-\frac{1}{4}g = -11 \longleftrightarrow -\frac{1}{4}g \times$ ______ $= -11 \times$ ______ $\longleftrightarrow g =$ ______

Multiply each side by the reciprocal of $-\frac{1}{4}$, which is -4.
Solution set is ______.

c. $-\frac{28}{3}q = 7 \longleftrightarrow -\frac{28}{3}q \times$ ______ $= 7 \times$ ______ $\longleftrightarrow q =$ ______

Multiply each side by the reciprocal of $-\frac{28}{3}$, which is ______.
Solution set is ______.

a. $-\frac{1}{5}$; $-\frac{1}{5}$; -4; $\{-4\}$ *b.* -4; -4; 44; {44}

c. $-\frac{3}{28}$; $-\frac{3}{28}$; $-\frac{3}{4}$; $-\frac{3}{28}$; $\left\{-\frac{3}{4}\right\}$

23 Solve the following equations by using multiplication:

a. $\frac{5}{6}y = 12 \longleftrightarrow \frac{5}{6}y \times$ ______ $= 12 \times$ ______ $\longleftrightarrow y =$ ______

Multiply each side by $\frac{6}{5}$. Solution set is ______.

b. $-\frac{8}{3}y = 2 \longleftrightarrow -\frac{8}{3}y \times$ ______ $= 2 \times$ ______ $\longleftrightarrow y =$ ______

Multiply each side by ______. Solution set is ______.

c. $7y = -11 \longleftrightarrow 7y \times$ ______ $= -11 \times$ ______ $\longleftrightarrow y =$ ______

Multiply each side by ______. Solution set is ______.

a. $\frac{6}{5}$; $\frac{6}{5}$; $\frac{72}{5}$; $\left\{\frac{72}{5}\right\}$ *b.* $-\frac{3}{8}$; $-\frac{3}{8}$; $-\frac{3}{4}$; $-\frac{3}{8}$; $\left\{-\frac{3}{4}\right\}$

c. $\frac{1}{7}$; $\frac{1}{7}$; $-\frac{11}{7}$; $\frac{1}{7}$; $\left\{-\frac{11}{7}\right\}$

24 Solve the following equations by using multiplication:

a. $-3k = 16 \longleftrightarrow -3k \times$ ______ $= 16 \times$ ______ $\longleftrightarrow k =$ ______
Multiply each side by ______. Solution set is ______.

b. $\frac{2}{3}z = 5 \longleftrightarrow \frac{2}{3}z \times$ ______ $= 5 \times$ ______ $\longleftrightarrow z =$ ______
Solution set is ______.

c. $-\frac{3}{7}p = -11 \longleftrightarrow -\frac{3}{7}p \times$ ______ $= -11 \times$ ______ $\longleftrightarrow p =$ ______
Solution set is ______.

---

a. $-\frac{1}{3}; -\frac{1}{3}; -\frac{16}{3}; -\frac{1}{3}; \left\{-\frac{16}{3}\right\}$ *b.* $\frac{3}{2}; \frac{3}{2}; \frac{15}{2}; \left\{\frac{15}{2}\right\}$

c. $-\frac{7}{3}; -\frac{7}{3}; \frac{77}{3}; \left\{\frac{77}{3}\right\}$

---

25 Solve the following equations:

a. $4z = -22 \longleftrightarrow 4z \times$ ______ $= -22 \times$ ______ $\longleftrightarrow z =$ ______
Solution set is ______.

b. $-\frac{5}{4}z = 30 \longleftrightarrow -\frac{5}{4}z \times$ ______ $= 30 \times$ ______ $\longleftrightarrow z =$ ______
Solution set is ______.

---

a. $\frac{1}{4}; \frac{1}{4}; -\frac{11}{2}; \left\{-\frac{11}{2}\right\}$ *b.* $-\frac{4}{5}; -\frac{4}{5}; -24; \{-24\}$

---

26 Solve the following equations:

a. $5z = -23 \longleftrightarrow 5z \times$ ______ $= -23 \times$ ______ $\longleftrightarrow z =$ ______
Solution set is ______.

b. $7z = -\frac{5}{2} \longleftrightarrow 7z \times$ ______ $= -\frac{5}{2} \times$ ______ $\longleftrightarrow z =$ ______
Solution set is ______.

---

a. $\frac{1}{5}; \frac{1}{5}; -\frac{23}{5}; \left\{-\frac{23}{5}\right\}$ *b.* $\frac{1}{7}; \frac{1}{7}; -\frac{5}{14}; \left\{-\frac{5}{14}\right\}$

---

27 Solve the following equations:

a. $-4z = \frac{7}{3} \longleftrightarrow -4z \times$ ______ $= \frac{7}{3} \times$ ______ $\longleftrightarrow z =$ ______

Solution set is ______.

b. $-\frac{3}{5}z = \frac{4}{7} \longleftrightarrow -\frac{3}{5}z \times$ ______ $= \frac{4}{7} \times$ ______ $\longleftrightarrow z =$ ______

Solution set is ______.

---

a. $-\frac{1}{4}; -\frac{1}{4}; -\frac{7}{12}; \left\{-\frac{7}{12}\right\}$ *b.* $-\frac{5}{3}; -\frac{5}{3}; -\frac{20}{21}; \left\{-\frac{20}{21}\right\}$

---

28 Solve the following equations:

a. $-6d = 21$

b. $-12d = \frac{7}{5}$

---

a. $\left\{-\frac{7}{2}\right\}$ *b.* $\left\{-\frac{7}{60}\right\}$

---

29 Solve the following equations:

a. $\frac{3}{11}q = -9$

b. $\frac{4}{5}b = -\frac{3}{11}$

---

a. $\{-33\}$ *b.* $\left\{-\frac{15}{44}\right\}$

---

30 Solve the following equations:

a. $\frac{5}{4}y = -\frac{3}{8}$

b. $-\frac{6}{7}z = -\frac{9}{10}$

---

a. $\left\{-\frac{3}{10}\right\}$ *b.* $\left\{\frac{21}{20}\right\}$

You have finished the program for Sec. 2-2. Now take the posttest to see how much you have learned.

POSTTEST 2-2

Convert the following equations to equivalent equations of the form $p = ?$:

1. $2p = -8 \longleftrightarrow p =$ _____
2. $-\frac{3}{11}p = \frac{5}{7} \longleftrightarrow p =$ _____
3. $-12p = -3 \longleftrightarrow p =$ _____

Give the solution sets of the following equations:

4. $-9k = \frac{10}{3}$
5. $-\frac{4}{5}y = -3$
6. $\frac{5}{9}z = 8$

When you have finished the posttest, correct it yourself, using the answers on page 134. Then check all the questions you got wrong on the list below:

1. _____
4. _____
2. _____
5. _____
3. _____
6. _____

If you compare your performance with that on Pretest 2-2, you will see how much you have improved. If you made more than one mistake on this posttest, review the material and take the posttest again. Don't move on to the next section until you can get all the questions right.

Section 2-3 Using the AT and the MT to Solve Equations

Do this section if you made a mistake in Questions 1 to 4 of Diagnostic Test 2. Begin by taking the pretest below, which will show you how much of Sec. 2-3 you already know and what you will be able to do by the time you have finished it.

PRETEST 2-3

Solve the following equations and check your answers.

1. $-3y + 5 = 7$
2. $4y - 6 = -11$
3. $-5 = 5 - 2y$
4. $k = -4k + 7$
5. $4z + 7 = 9z$
6. $3z - 11 = -5z$

When you have finished the pretest, correct it yourself using the answers on page 134. Then check all the questions you got wrong on the list below:

1. ______ 3. ______ 5. ______

2. ______ 4. ______ 6. ______

If you made any mistakes (even if it was only *one* mistake), read the introduction and go through the program.

If you made no mistakes at all, skip Sec. 2-3, because you already know its material, and go on to Sec. 2-4.

Introduction

Adding the same real number to both sides of an equation is called the *addition transformation* (AT). Multiplying both sides of an equation by the same real number is called the *multiplication transformation* (MT). You learn how to use combinations of both transformations to solve equations of the general type $ax + b = c$. You are shown that the addition transformation is valid for multiples of the variable as well as for real numbers, and you use this property to solve equations of the general type $ax = bx + c$. You will also learn how to check a solution by substitution in the original equation.

FRAMES FOR SECTION 2-3

1 Convert the following equations to equivalent equations of the form $y = ?$:

a. $y + 2 = 4 \longleftrightarrow y =$ ______

b. $y - 3 = 8 \longleftrightarrow y =$ ______

c. $3y = 12 \longleftrightarrow y =$ ______

d. $\dfrac{y}{8} = 2 \longleftrightarrow y =$ ______

---

a. 2 *b.* 11 *c.* 4 *d.* 16

---

2 *a.* In Frame 1*a* you had to add ______ to both sides.
b. In 1*b* you added ______.
c. In 1*c* you multiplied both sides by ______.
d. In 1*d* you multiplied both sides by ______.

---

a. -2 *b.* 3 *c.* $\frac{1}{3}$ *d.* 8

---

3 Adding the same number to both sides of an equation is called an addition transformation; and multiplying both sides of an equation by the same number is called a multiplication transformation.

a. To convert $3y = 5$ to $y =$ ______ you use the ______________ transformation with $\dfrac{1}{3}$.

b. To convert $y + 5 = 7$ to $y =$ ______ you use the ______________ transformation with -5.

---

a. $\frac{5}{3}$; multiplication *b.* 2; addition

---

4 An equation of the form $3y + 7 = 5$ can be solved by using both transformations:

a. Using the ______________ transformation with -7

$$3y + 7 = 5 \longleftrightarrow 3y = -2$$

b. Using the ______________ transformation with $\dfrac{1}{3}$

$$3y = -2 \longleftrightarrow y = \text{\_\_\_\_\_\_}$$

---

a. addition *b.* multiplication; $-\frac{2}{3}$

---

5 Convert the following equations to the form $k = ?$ by using the AT followed by the MT:

		AT		MT	
a.	$3k + 2 = 4$	$\longleftrightarrow$	$3k =$ ______	$\longleftrightarrow$	$k =$ ______
b.	$6 = 5k - 1$	$\longleftrightarrow$	______ $= 5k$	$\longleftrightarrow$	______ $= k$
c.	$5k - 1 = 7$	$\longleftrightarrow$	______ $=$ ______	$\longleftrightarrow$	$k =$ ______
d.	$2 = 4k + 6$	$\longleftrightarrow$	______ $=$ ______	$\longleftrightarrow$	______ $= k$

---

a. $3k = 2 \longleftrightarrow k = \frac{2}{3}$ *b.* $7 = 5k \longleftrightarrow \frac{7}{5} = k$
c. $5k = 8 \longleftrightarrow k = \frac{8}{5}$ *d.* $-4 = 4k \longleftrightarrow -1 = k$

---

6 Convert the following equations to the form $k = ?$

		AT		MT	
a.	$6k - 2 = -3$	$\longleftrightarrow$	______ $=$ ______	$\longleftrightarrow$	$k =$ ______
b.	$-7 = -3 + 2k$	$\longleftrightarrow$	______ $=$ ______	$\longleftrightarrow$	______ $= k$
c.	$8 = 11 - 6k$	$\longleftrightarrow$	______ $=$ ______	$\longleftrightarrow$	______ $= k$
d.	$-5k + 7 = 12$	$\longleftrightarrow$	______ $=$ ______	$\longleftrightarrow$	$k =$ ______

---

a. $6k = -1 \longleftrightarrow k = -\frac{1}{6}$ *b.* $-4 = 2k \longleftrightarrow -2 = k$
c. $-3 = -6k \longleftrightarrow \frac{1}{2} = k$ *d.* $-5k = 5 \longleftrightarrow k = -1$

---

7 Solve the following equations; i.e. give their solution sets:

a. $2y - 3 = -7$
b. $-5y + 4 = 1$
c. $4 = 3y - 2$
d. $-8 = 2 - 3y$

---

a. $\{-2\}$ *b.* $\{\frac{3}{5}\}$ *c.* $\{2\}$ *d.* $\{\frac{10}{3}\}$

---

8 *a.* 1 (is/is not) ________ a solution of $2y + 4 = 8$, because $(2 \times 1) + 4 = 8$ is a (true/false) ________ statement.

b. 2 (is/is not) ________ a solution of $2y + 4 = 8$, because $(2 \times 2) + 4 = 8$ is a (true/false) ________ statement.

c. 3 (is/is not) ________ a solution of $2y + 4 = 8$, because $(2 \times 3) + 4 = 8$ is a (true/false) ________ statement.

---

a. is not; false *b.* is; true *c.* is not; false

---

9 In Frame 8 you found whether 1, 2, and 3 are solutions of $2y + 4 = 8$ by the process we call *substitution*.

a. In 8*a* you substituted ______ for y in the open sentence $2y + 4 = 8$ to obtain the statement $(2 \times 1) + 4 = 8$.

b. In 8*b* you substituted ______ for y in the open sentence $2y + 4 = 8$ to obtain the statement ____________.

c. Substituting 4 for y in the open sentence $2y + 4 = 8$ would give you the statement ____________, which is (true/false) ________.

---

a. 1 *b.* 2; $(2 \times 2) + 4 = 8$ *c.* $(2 \times 4) + 4 = 8$; false

---

10 *a.* Substituting 5 for y in the sentence $3y - 2 = 13$ gives the statement ____________, which is (true/false) ______.

b. Is 5 a solution of $3y - 2 = 13$?

c. How could you find out if 6 is a solution of $2y + 3 = 15$?

d. Is 6 a solution of $2y + 3 = 15$?

---

a. $(3 \times 5) - 2 = 13$; true *b.* yes
c. by substituting 6 for y in the sentence $2y + 3 = 15$ to get a statement and checking if the statement is true or false *d.* yes

---

11 When you have solved an equation, you should always check your answer by substituting it in the equation and seeing if it gives a true statement.

a. In Frame 7*a* we deduced that the solution set of $2y - 3 = -7$ was $\{-2\}$. Substituting -2 for y we get the statement ____________, which is (true/false) ______.

b. In Frame 7*c* we deduced that the solution set of $4 = 3y - 2$ was $\{2\}$. Check this answer by substitution.

---

a. $2(-2) - 3 = -7$; true

b. Substituting 2 for y we get the statement $4 = (3 \times 2) - 2$, which is true. Therefore 2 is a solution of $4 = 3y - 2$.

---

12 Solve the equation $3y + 5 = 2$ by using the AT followed by the MT. Then check your answer by substitution. Show all your work.

$$3y + 5 = 2$$

$\updownarrow$ AT with -5

$$3y = -3$$

$\updownarrow$ MT with $\frac{1}{3}$

$$y = -1$$

$\{-1\}$ is the solution set.
Substituting -1 for y, we get $3(-1) + 5 = 2$, which is true.

---

13 Solve the equation $4y + 9 = 3$ and check your answer by substitution.

$$4y + 9 = 3$$
$$4y = 3 - 9 = -6$$

$\left\{-\frac{3}{2}\right\}$ is the solution set.

CHECK

$$4\left(-\frac{3}{2}\right) + 9 = 3$$

$$-\frac{12}{2} + 9 = 3$$

$$-6 + 9 = 3 \qquad \text{True}$$

---

14 *a.* If the domain of y is $\{1, 2, 3, 4\}$, we obtain ______ (how many?) statements from the open sentence $2y = y + 1$.

b. Which of these statements is true? (Write out the statement in full.)

---

a. four *b.* $2 \times 1 = 1 + 1$ is true

---

15 What are the solution sets of the following open sentences when the domain of y is $\{1, 2, 3, 4\}$?

a. $y = 1$
b. $2y = y + 1$
c. $3y = 2y + 1$
d. $4y = 3y + 1$

---

a. $\{1\}$ *b.* $\{1\}$ *c.* $\{1\}$ *d.* $\{1\}$

---

16 What are the solution sets of the following open sentences when the domain of y is the set of real numbers? Check your answers by substitution.

a. $y = 3$
b. $2y = y + 3$
c. $3y = 2y + 3$
d. $4y = 3y + 3$

---

a. $\{3\}$; $3 = 3$ is true
b. $\{3\}$; $2 \times 3 = 3 + 3$ is true
c. $\{3\}$; $3 \times 3 = (2 \times 3) + 3$ is true
d. $\{3\}$; $4 \times 3 = (3 \times 3) + 3$ is true

---

17 Which of the following pairs of equations are equivalent?

a. $y = 5$ and $3y = 2y + 5$
b. $y = -2$ and $4y = 3y - 2$
c. $y = 3$ and $-2y = -3y + 3$

---

all of them

---

18 *a.* The second equation in Frame 17*a* could be obtained by adding ______ to both sides of the first equation.

b. The second equation in 17*b* could be obtained by adding ______ to both sides of the first equation.

c. The second equation in 17*c* could be obtained by adding ______ to both sides of the first equation.

---

a. $2y$ *b.* $3y$ *c.* $-3y$

---

19 If we add any multiple of the variable to both sides of an equation, we obtain an equivalent equation:

$$2y = 5 \longleftrightarrow 2y + ny = \underline{\hspace{6em}}$$

where n is any real number. Therefore the addition transformation applies to any real number and to ____________.

$2y + ny = 5 + ny$; any multiple of the variable

---

20 To solve equations of the form $7y = 5y + 3$:

a. First use the ______ transformation with $-5y$ and simplify the resultant expression:

$$7y = 5y + 3 \longleftrightarrow 2y = 3$$

b. Then use the ____________ transformation with $\frac{1}{2}$:

$$2y = 3 \longleftrightarrow y = \underline{\hspace{3em}}$$

---

a. addition *b.* multiplication; $\frac{3}{2}$

---

21 Convert the following equations to the form $k = ?$ by using the addition transformation followed by the multiplication transformation:

		AT		MT	
a.	$8k = 3k + 7$	$\longleftrightarrow$	______ $=$ 7	$\longleftrightarrow$	$k =$ ______
b.	$-4k + 9 = 3k$	$\longleftrightarrow$	9 $=$ ______	$\longleftrightarrow$	______ $= k$
c.	$-2k - 4 = 5k$	$\longleftrightarrow$	______ $=$ ______	$\longleftrightarrow$	______ $= k$
d.	$k = 5k + 5$	$\longleftrightarrow$	______ $=$ ______	$\longleftrightarrow$	$k =$ ______

---

a. $5k = 7 \longleftrightarrow k = \frac{7}{5}$ *b.* $9 = 7k \longleftrightarrow \frac{9}{7} = k$

c. $-4 = 7k \longleftrightarrow -\frac{4}{7} = k$ *d.* $-4k = 5 \longleftrightarrow k = -\frac{5}{4}$

---

22 Convert the following equations to the form $k = ?$:

		AT		MT	
a.	$5k = 2k + 2$	$\longleftrightarrow$	______ $=$ ______	$\longleftrightarrow$	$k =$ ______
b.	$-3k = 2 - 2k$	$\longleftrightarrow$	______ $=$ ______	$\longleftrightarrow$	$k =$ ______
c.	$7k + 3 = 3k$	$\longleftrightarrow$	______ $=$ ______	$\longleftrightarrow$	______ $= k$
d.	$2k - 5 = 6k$	$\longleftrightarrow$	______ $=$ ______	$\longleftrightarrow$	______ $= k$

---

a. $3k = 2 \longleftrightarrow k = \frac{2}{3}$ *b.* $-k = 2 \longleftrightarrow k = -2$

c. $3 = -4k \longleftrightarrow -\frac{3}{4} = k$ *d.* $-5 = 4k \longleftrightarrow -\frac{5}{4} = k$

---

23 Solve the following equations and then check your answers by substitution:

a. $3k - 4 = k$

b. $6k + 7 = -2k$

c. $-3k = 2k - 6$

---

a. $\{2\}$

CHECK

$(3 \times 2) - 4 = 2$ True

b. $\left\{-\frac{7}{8}\right\}$

CHECK

$$6\left(-\frac{7}{8}\right) + 7 = -2\left(-\frac{7}{8}\right)$$

$$-\frac{42}{8} + 7 = \frac{14}{8}$$

$$-\frac{42}{8} + \frac{56}{8} = \frac{14}{8}$$ True

c. $\left\{\frac{6}{5}\right\}$

CHECK

$$-3\left(\frac{6}{5}\right) = 2\left(\frac{6}{5}\right) - 6$$

$$-\frac{18}{5} = \frac{12}{5} - 6$$

$$-\frac{18}{5} = \frac{12}{5} - \frac{30}{5}$$ True

---

24 Solve the following equations and then check your answers by substitution:

a. $k = -4k - 7$
b. $5k + 2 = 6$
c. $-5k = 7k + 2$

---

a. $\left\{-\frac{7}{5}\right\}$

CHECK

$$-\frac{7}{5} = -4\left(-\frac{7}{5}\right) - 7$$

$$-\frac{7}{5} = \frac{28}{5} - 7$$

$$-\frac{7}{5} = \frac{28}{5} - \frac{35}{5} \qquad \text{True}$$

b. $\left\{\frac{4}{5}\right\}$

CHECK

$$5\left(\frac{4}{5}\right) + 2 = 6$$

$$4 + 2 = 6 \qquad \text{True}$$

c. $\left\{-\frac{1}{6}\right\}$

CHECK

$$-5\left(-\frac{1}{6}\right) = 7\left(-\frac{1}{6}\right) + 2$$

$$\frac{5}{6} = -\frac{7}{6} + 2$$

$$\frac{5}{6} = -\frac{7}{6} + \frac{12}{6} \qquad \text{True}$$

You have finished the program for Sec. 2-3. Now take the posttest to see how much you have learned.

POSTTEST 2-3

Solve the following equations and check your answers:

1. $-2y + 8 = 5$
2. $3y - 5 = -7$
3. $-2 = 8 - 5y$
4. $k = -7k + 4$
5. $5z + 9 = 2z$
6. $6z - 8 = -3z$

When you have finished the posttest, correct it yourself using the answers on page 134. Then check all the questions you got wrong on the list below:

1. ______ 3. ______ 5. ______

2. ______ 4. ______ 6. ______

If you compare your performance with that on Pretest 2-3, you will see how much you have improved. If you made more than one mistake on this posttest, review the material and take the pretest again. Don't move on to the next section until you can get all the questions right.

Now proceed to Sec. 2-4.

Section 2-4
Solving Equations of the Form $ay + b = cy + d$

Do this section if you made a mistake in Questions 1 to 4 of Diagnostic Test 2. Begin by taking the pretest below, which will show you how much of Sec. 2-4 you already know and what you will be able to do by the time you have finished it.

PRETEST 2-4

Solve the following equations and check your answers:

1. $7y + 8 - 3y - 2 = 5y - 8 + 3$
2. $6y + 7 - 13y = 4y - 12 + 8$
3. $2 - 3y + 8 = 4y - 2 - 9y$
4. $4y + 3 - 9 = 11 - 3y + 8y$
5. $5 + 5y - 10 = 2y + 8 - 6y$
6. $3 + 3y - 10y = 7 - y + 1$

When you have finished the pretest, correct it yourself using the answers on page 134. Then check all the questions you got wrong on the list below:

1. ______ 3. ______ 5. ______

2. ______ 4. ______ 6. ______

If you made any mistakes (even if it was only *one* mistake) read the introduction and go through the program.

If you made no mistakes at all, skip Sec. 2-4, because you already know its material, and go on to the next section whose number you circled on page 62. If no further numbers were circled, take Final Test 2 (page 132).

Introduction

Section 2-4 shows you how to extend the methods of the previous sections to solve the solution of equations with more than three terms, but without any fractions or parentheses.

FRAMES FOR SECTION 2-4

1 To solve equations of the type

$$5y + 6 + 2y = 3 + y + 1$$

apply several addition transformations in succession until all the terms containing variables are on one side of the equation and all the terms without variables are on the other side:

$$5y + 6 + 2y = 3 + y + 1$$

$\updownarrow$ Adding ______ to both sides

$$5y + 6 + 2y + (-y) = 3 + 1$$

$\updownarrow$ Adding ______ to both sides

$$5y + 2y + (-y) = 3 + 1 + (-6)$$

$-y$ $\quad -6$

---

2 Then simplify the expressions on either side

$$5y + 2y + (-y) = 3 + 1 + (-6)$$

$\updownarrow$

______ = ______

and use the multiplication transformation to obtain the solution set, ______.

$6y = -2$ $\quad \{-\frac{1}{3}\}$

---

3 Check your answer by substituting $-\dfrac{1}{3}$ for y in the original equation:

$$5y + 6 + 2y = 3 + y + 1$$

$$5\left(-\frac{1}{3}\right) + 6 + 2\left(-\frac{1}{3}\right) = 3 + \left(-\frac{1}{3}\right) + 1$$

$$-\frac{5}{3} + 6 - \frac{2}{3} = 3 - \frac{1}{3} + 1$$

$$6 - \frac{7}{3} = 4 - \frac{1}{3}$$

$$\frac{18 - 7}{3} = \frac{12 - 1}{3}$$

$$\frac{11}{3} = \frac{11}{3} \qquad \text{True}$$

---

4 To solve $7y + 8 = 4y + 2y + 3$ apply the addition transformation several times to get all the terms containing the variable on one side and all the terms without variables on the other side.

$7y + 8 = 4y + 2y + 3$

↕ AT using $-7y$

____

↕ AT using -3

____

$8 = 4y + 2y + 3 + (-7y)$

$8 + (-3) = 4y + 2y + (-7y)$

---

5 $8 + (-3) = 4y + 2y + (-7y)$

↕ Simplify both expressions

______ = ______

↕ MT with -1

______ $= y$

Solution set is ______. Check this answer by substitution in the original equation. The statement ____________ is (true/false) ______.

$$5 = -y$$
$$-5 = y$$
$$\{-5\}$$

CHECK

$$7(-5) + 8 = 4(-5) + 2(-5) + 3$$
$$-35 + 8 = -20 - 10 + 3$$
$$-27 = -27 \quad \text{True}$$

6 $$6y + 4 + 2y = 5y + 3 + 2$$

↕ AT with $-5y$ and with -4

____________________ = ____________________

(only terms containing the variable) (only terms not containing the variable)

↕ Simplify both expressions

______ = ______

↕ MT with $\frac{1}{3}$

$y =$ ______

Solution set is ______. Check your answer by substitution.

$$6y + 2y + (-5y) = 3 + 2 + (-4)$$
$$3y = 1$$
$$y = \tfrac{1}{3}$$
$$\{\tfrac{1}{3}\}$$

CHECK

$$6\left(\frac{1}{3}\right) + 4 + 2\left(\frac{1}{3}\right) = 5\left(\frac{1}{3}\right) + 3 + 2$$
$$2 + 4 + \frac{2}{3} = \frac{5}{3} + 3 + 2$$
$$\frac{20}{3} = \frac{20}{3} \quad \text{True}$$

7 $$4y - 5 + 3 = -7y + 2y + 6$$

$\updownarrow$ AT with $-4y$ and -6

____________________ = ____________________

(only terms *not* containing the variable) (only terms containing the variable)

$\updownarrow$ Simplify

______ = ______

$\updownarrow$ MT

______ $= y$

Solution set is ________________. Check your answer.

$$-5 + 3 + (-6) = -7y + 2y + (-4y)$$

$$-8 = -9y$$

$$\frac{8}{9} = y$$

$$\left\{\frac{8}{9}\right\}$$

CHECK

$$4\left(\frac{8}{9}\right) - 5 + 3 = -7\left(\frac{8}{9}\right) + 2\left(\frac{8}{9}\right) + 6$$

$$\frac{32}{9} - 2 = -\frac{56}{9} + \frac{16}{9} + 6$$

$$\frac{32}{9} - \frac{18}{9} = -\frac{40}{9} + \frac{54}{9}$$

$$\frac{14}{9} = \frac{14}{9} \quad \text{True}$$

---

8 $$4y + 3y - 1 = 2y - 8 + 3$$

$\updownarrow$ AT with _____ and with _____

____________________ = ____________________

(only terms containing the variable) (only terms *not* containing the variable)

$\updownarrow$ Simplify

_____ = _____

$\updownarrow$ MT

_____ = _____

Solution set is _____. Check your answer.

$-2y$; 1

$$4y + 3y + (-2y) = -8 + 3 + 1$$

$$5y = -4$$

$$y = -\frac{4}{5}$$

$$\left\{-\frac{4}{5}\right\}$$

CHECK

$$4\left(-\frac{4}{5}\right) + 3\left(-\frac{4}{5}\right) - 1 = 2\left(-\frac{4}{5}\right) - 8 + 3$$

$$-\frac{16}{5} - \frac{12}{5} - \frac{5}{5} = -\frac{8}{5} - \frac{40}{5} + \frac{15}{5}$$

$$-\frac{33}{5} = -\frac{33}{5} \quad \text{True}$$

---

9 $$3y - 5y + 2 + 4 = 6y - 2y + 8$$

$\updownarrow$ AT with ______ and ______ and ______

______________________ = ______________________

(only terms *not* containing the variable) (only terms containing the variable)

$\updownarrow$ Simplify

______ = ______

$\updownarrow$ MT

______ = ______

Solution set is ______. Check your answer.

$-3y$; $5y$; -8

$$2 + 4 + (-8) = 6y - 2y + (-3y) + 5y$$

$$-2 = 6y$$

$$-\frac{1}{3} = y$$

$$\left\{-\frac{1}{3}\right\}$$

CHECK

$$3\left(-\frac{1}{3}\right) - 5\left(-\frac{1}{3}\right) + 2 + 4 = 6\left(-\frac{1}{3}\right) - 2\left(-\frac{1}{3}\right) + 8$$

$$-1 + \frac{5}{3} + 2 + 4 = -2 + \frac{2}{3} + 8$$

$$6\tfrac{2}{3} = 6\tfrac{2}{3} \qquad \text{True}$$

---

10 Solve $10y + 6 - 2y = 3y + 8 - 4$ and check your answer. Show all your work.

$$10y + 6 - 2y = 3y + 8 - 4$$
$$10y - 2y + (-3y) = 8 - 4 + (-6)$$
$$5y = -2$$
$$y = -\frac{2}{5}$$

$\left\{-\frac{2}{5}\right\}$

CHECK

$$10\left(-\frac{2}{5}\right) + 6 - 2\left(-\frac{2}{5}\right) = 3\left(-\frac{2}{5}\right) + 8 - 4$$
$$-4 + 6 + \frac{4}{5} = -\frac{6}{5} + 8 - 4$$
$$2 + \frac{4}{5} = -\frac{6}{5} + 4$$
$$\frac{14}{5} = \frac{-6 + 20}{5} \quad \text{True}$$

11 Solve $8y + 3 - y = 4y - 5 + 7$ and check your answer. Show all your work.

$$8y - y + (-4y) = -5 + 7 + (-3)$$
$$3y = -1$$
$$y = -\frac{1}{3}$$

$\left\{-\frac{1}{3}\right\}$

CHECK

$$8\left(-\frac{1}{3}\right) + 3 - \left(-\frac{1}{3}\right) = 4\left(-\frac{1}{3}\right) - 5 + 7$$
$$-\frac{8}{3} + 3 + \frac{1}{3} = -\frac{4}{3} + 2$$
$$3 - \frac{7}{3} = 2 - \frac{4}{3}$$
$$\frac{2}{3} = \frac{2}{3} \quad \text{True}$$

12 Solve $4y + 2y - 4 + 2 = 8y - 5 + 7 - y$ and check your answer.

$\{-4\}$

CHECK

$$\begin{aligned} 4(-4) + 2(-4) - 4 + 2 &= 8(-4) - 5 + 7 - (-4) \\ -16 - 8 - 4 + 2 &= -32 - 5 + 7 + 4 \\ -28 + 2 &= -37 + 11 \\ -26 &= -26 \qquad \text{True} \end{aligned}$$

---

13 Solve $8y + 5 - 3y + 2 = 6y + 8 - 7$ and check your answer.

$\{6\}$

CHECK

$$\begin{aligned} 8(6) + 5 - 3(6) + 2 &= 6(6) + 8 - 7 \\ 48 + 5 - 18 + 2 &= 36 + 8 - 7 \\ 55 - 18 &= 44 - 7 \\ 37 &= 37 \qquad \text{True} \end{aligned}$$

You have finished the program for Sec. 2-4. Now take the posttest to see how much you have learned.

POSTTEST 2-4

Solve the following equations and check your answers:

1. $5y + 9 - 6y - 3 = 7y - 4 + 18$
2. $3y + 4 - 10y = 6y - 5 + 8$
3. $4 - 6y + 9 = 5y - 1 - 7y$
4. $3y + 2 - 8 = 11 - 4y + 9y$
5. $6 + 3y - 5 = 3y + 9 - 4y$
6. $8 + 8y - 3y = 5 - y + 1$

When you have finished the posttest, correct it yourself using the answers on page 134. Then check all the questions you got wrong on the list below:

1. ______ 3. ______ 5. ______

2. ______ 4. ______ 6. ______

If you compare your performance with that on Pretest 2-4, you will see how much you have improved. If you made more than one mistake on this posttest, review the material and take the pretest again. Don't move on to the next section until you can get all the questions right.

Now proceed to the next section whose number you circled on page 62. If no other sections were circled, take Final Test 2 (page 132).

Section 2-5
Solving Equations Containing Parentheses

Do this section if you made a mistake in Question 5 or 6 of Diagnostic Test 2. Begin by taking the pretest below, which will show you how much of Sec. 2-5 you already know and what you will be able to do by the time you have finished it.

PRETEST 2-5

Find the solution set of the following equations and check your answers:

1. $3(k - 2) = 4(3 - 2k)$
2. $5(3 - 4k) = 2(k + 5)$
3. $2(3y - 2) - 6(2y + 7) = -3(y + 8)$
4. $4(3 - 7y) - 2(4 - 3y) = 2(3y + 8)$
5. $5(2y - 5) = 4(3y - 7) - 2(2y + 9)$
6. $-2(y - 6) = 3(5y - 8) - (y - 3)$

When you have finished the pretest, correct it yourself using the answers on page 135. Then check all the questions you got wrong on the list below:

1. ______ 3. ______ 5. ______

2. ______ 4. ______ 6. ______

If you made any mistakes (even if it was only *one* mistake), read the introduction and go through the program.

If you made no mistakes at all, skip Sec. 2-5, because you already know its material, and go on to the next section whose number you circled on page 62. If no further numbers were circled, take Final Test 2 (page 132).

Introduction

Now you will learn how to extend the methods of the previous sections to equations containing parentheses but no fractions. You begin by using the distributive law to remove parentheses.

FRAMES FOR SECTION 2-5

1 By the distributive law: $3(4y + 7) = 3(4y) + 3(7) = 12y + 21$.

a. $4(5y + 9) = 4(5y) + 4(9) =$ _____ + _____

b. $5(6y - 11) = 5(6y) - 5(11) =$ _____ − _____

---

a. $20y + 36$ *b.* $30y - 55$

---

2 *a.* $6(7y - 9) = 6(7y) - 6(9) =$ _____ − _____

b. $8(4y + 3) =$ _____ + _____

c. $5(2y - 7) =$ _____ − _____

---

a. $42y - 54$ *b.* $32y + 24$ *c.* $10y - 35$

---

3 *a.* $2(5y + 6) =$ _____ + _____

b. $3(2y - 5) =$ _____ − _____

c. $8(5 - 3y) =$ _____ − _____

---

a. $10y + 12$ *b.* $6y - 15$ *c.* $40 - 24y$

---

4 Expressions containing parentheses can often be simplified by using the distributive law to remove the parentheses and then combining terms.

a. $4(5 + 6y) + 3(2y + 5) =$ _____ + _____ + _____ + _____ =

_____

b. $2(4 - 3y) + 3(4y - 6) =$ _____ − _____ + _____ − _____ =

_____

---

a. $20 + 24y + 6y + 15 = 30y + 35$
b. $8 - 6y + 12y - 18 = 6y - 10$

---

5 Simplify the following expressions:

a. $2(6 - 5y) + 5(3y - 7)$
b. $8(2 - 5y) + 3(y - 2)$
c. $3(1 - 6y) + 2(y - 5)$

---

a. $5y - 23$ *b.* $-37y + 10$ *c.* $-16y - 7$

---

6 Simplify the following expressions:

a. $6(1 - 3y) + 4(y - 8)$
b. $3(5y - 4) + 5(2 - 3y)$
c. $4(3 - 2y) + 6(y - 2)$

---

a. $-14y - 26$ *b.* -2 *c.* $-2y$

---

7 *a.* $7(2y + 3) - 3(y + 2) = 7(2y + 3) + (-3)(y + 2)$
$= \_\_\_\_ + \_\_\_\_ + (-3)y + (-3)(2)$
$= \_\_\_\_ + \_\_\_\_ - \_\_\_\_ - \_\_\_\_$
$= \_\_\_\_\_\_$

b. $4(y + 5) - 2(3y - 7) = 4(y + 5) + (-2)(3y - 7)$
$= \_\_\_\_ + \_\_\_\_ + (-2)(3y) - (-2)(7)$
$= \_\_\_\_ + \_\_\_\_ - \_\_\_\_ + \_\_\_\_$
$= \_\_\_\_\_\_$

---

a. $14y + 21$; $14y + 21 - 3y - 6$; $11y + 15$
b. $4y + 20$; $4y + 20 - 6y + 14$; $-2y + 34$

---

8 *a.* $3(y + 2) - 4(2y - 7) = \_\_\_\_ + \_\_\_\_ + (-4)(2y) - (-4)(7)$
$= \_\_\_\_ + \_\_\_\_ - \_\_\_\_ + \_\_\_\_$
$= \_\_\_\_\_\_$

b. $5(y - 1) - 2(3y + 4) = \_\_\_\_ - \_\_\_\_ + (-2)(3y) + (-2)(4)$
$= \_\_\_\_ - \_\_\_\_ - \_\_\_\_ - \_\_\_\_$
$= \_\_\_\_\_\_$

---

a. $3y + 6$; $3y + 6 - 8y + 28$; $-5y + 34$
b. $5y - 5$; $5y - 5 - 6y - 8$; $-y - 13$

---

9 *a.* $5(2y - 3) - 3(3y - 4) = \_\_\_\_\_\_ + (-3)(3y) - (-3)(4)$
$= \_\_\_\_\_\_\_\_\_\_$
$= \_\_\_\_\_\_$

b. $4(3 - 4y) - 5(2y + 3) = \_\_\_\_\_\_ + (-5)(2y) + (-5)(3)$
$= \_\_\_\_\_\_\_\_\_\_$
$= \_\_\_\_\_\_$

---

a. $10y - 15$; $10y - 15 - 9y + 12$; $y - 3$
b. $12 - 16y$; $12 - 16y - 10y - 15$; $-26y - 3$

---

10 Put a + or a − in front of every term. (Look back at previous frames if you need to.)

a. $3(6y - 5) + 4(2y - 5) =$ _____$18y$ _____15 _____$8y$ _____20

b. $3(6y + 5) - 4(2y - 5) =$ _____$18y$ _____15 _____$8y$ _____20

c. $3(6y - 5) - 4(2y + 5) =$ _____$18y$ _____15 _____$8y$ _____20

d. $-3(6y + 5) + 4(2y + 5) =$ _____$18y$ _____15 _____$8y$ _____20

---

a. $+18y - 15 + 8y - 20$ *b.* $+18y + 15 - 8y + 20$
c. $+18y - 15 - 8y - 20$ *d.* $-18y - 15 + 8y + 20$

---

11 Put a + or a − in front of every term.

a. $2(y + 3) - 3(y - 7) =$ _____$2y$ _____6 _____$3y$ _____21

b. $-2(y + 3) - 3(y + 7) =$ _____$2y$ _____6 _____$3y$ _____21

c. $-2(y - 3) + 3(y - 7) =$ _____$2y$ _____6 _____$3y$ _____21

d. $2(y - 3) - 3(y + 7) =$ _____$2y$ _____6 _____$3y$ _____21

---

a. $+2y + 6 - 3y + 21$ *b.* $-2y - 6 - 3y - 21$
c. $-2y + 6 + 3y - 21$ *d.* $+2y - 6 - 3y - 21$

---

12 Simplify the following expressions:

a. $3(y - 5) + 6(2y - 3)$
b. $2(y - 7) - 4(y + 3)$
c. $4(y + 1) - 3(y - 2)$

---

a. $3y - 15 + 12y - 18 = 15y - 33$
b. $2y - 14 - 4y - 12 = -2y - 26$
c. $4y + 4 - 3y + 6 = y + 10$

---

13 Simplify the following expressions:

a. $-2(y + 6) + 3(7 - 2y)$
b. $-4(2y - 3) - 7(2 - 3y)$
c. $6(4 - 3y) - 8(2y + 5)$

---

a. $-8y + 9$ *b.* $13y - 2$ *c.* $-34y - 16$

---

14 To solve equations containing parentheses, first remove the parentheses and then solve the equation in the usual manner.

$$3(k - 7) = 4(2 - k) \longleftrightarrow 3k - 21 = \_\_\_\_\_\_$$

which can be solved to give $k =$ ______.

$8 - 4k$ $\quad \frac{29}{7}$

---

15 Check your answer by the method indicated:

$$3\left(\frac{29}{7} - 7\right) = 4\left(2 - \frac{29}{7}\right)$$

$$3\left(\frac{29 - \_\_}{7}\right) = 4\left(\frac{\_\_ - 29}{7}\right)$$

$$\frac{3(-\_\_)}{7} = \frac{4(-\_\_)}{7}$$

$$-\frac{\_\_}{7} = -\frac{\_\_}{7}$$

$$3\left(\frac{29 - 49}{7}\right) = 4\left(\frac{14 - 29}{7}\right)$$

$$\frac{3(-20)}{7} = \frac{4(-15)}{7}$$

$$-\frac{60}{7} = -\frac{60}{7} \quad \text{True}$$

---

16 Solve the equation $6(k - 3) = 5(2 - 3k)$ and check your answer by the method used in Frame 15. (Refer to Frame 15 if necessary.)

$\left\{\frac{4}{3}\right\}$

CHECK

$$6\left(\frac{4}{3} - 3\right) = 5(2 - 4)$$

$$\overset{2}{\not 6}\left(\frac{4 - 9}{\underset{1}{\not 3}}\right) = 5(-2)$$

$$2(-5) = 5(-2) \quad \text{True}$$

---

17 Solve the equation $3(z + 2) - 5(3z + 4) = -2(z + 3)$ and check your answer by the method used in Frame 15. (Refer to Frame 15 if necessary.)

$3z + 6 - 15z - 20 = -2z - 6$

$\left\{-\frac{4}{5}\right\}$

CHECK

$$3\left(-\frac{4}{5} + 2\right) - 5\left(-\frac{12}{5} + 4\right) = -2\left(-\frac{4}{5} + 3\right)$$

$$3\left(\frac{-4 + 10}{5}\right) - 5\left(\frac{-12 + 20}{5}\right) = -2\left(\frac{-4 + 15}{5}\right)$$

$$\frac{3(6)}{5} - \frac{5(8)}{5} = \frac{-2(11)}{5}$$

$$\frac{18}{5} - \frac{40}{5} = \frac{-22}{5} \qquad \text{True}$$

---

18 Solve the equation $4(2y + 5) - 3(2 - 3y) = 5(y + 2)$ and check your answer.

$\left\{-\frac{1}{3}\right\}$

CHECK

$$4\left(-\frac{2}{3} + 5\right) - 3(2 + 1) = 5\left(-\frac{1}{3} + 2\right)$$

$$4\left(\frac{-2 + 15}{3}\right) - 3(3) = 5\left(\frac{-1 + 6}{3}\right)$$

$$\frac{4(13)}{3} - 9 = \frac{5(5)}{3}$$

$$\frac{52}{3} - \frac{27}{3} = \frac{25}{3} \qquad \text{True}$$

You have finished the program for Sec. 2-5. Now take the posttest to see how much you have learned.

POSTTEST 2-5

Find the solution sets of the following equations and check your answers:

1. $8(k - 3) = 3(2 - 3k)$
2. $4(5 - 2k) = 5(k + 3)$
3. $5(2z - 9) - 4(3z + 2) = -6(z + 7)$
4. $3(2 - 5y) - 6(3 - 4y) = 7(2y + 9)$
5. $2(7y - 3) = 5(2y - 5) - 3(4y + 7)$
6. $-3(y - 4) = 5(2y - 9) - (y - 5)$

When you have finished the posttest, correct it yourself using the answers on page 135. Then check all the questions you got wrong on the list below:

1. ______ 3. ______ 5. ______

2. ______ 4. ______ 6. ______

If you compare your performance with that on Pretest 2-5, you will see how much you have improved. If you made more than one mistake on this posttest, review the material and take the pretest again. Don't move on to the next section until you can get all the questions right.

Now proceed to the next section whose number you circled on page 62. If no other sections were circled, take Final Test 2 (page 132).

Section 2-6
Solving Equations Containing Fractions

Do this section if you made a mistake in Questions 7 to 9 of Diagnostic Test 2. Begin by taking the pretest below, which will show you how much of Sec. 2-6 you already know and what you will be able to do by the time you have finished it.

PRETEST 2-6

Solve the following equations and check your answers.

1. $\frac{3y}{4} - \frac{y}{2} + \frac{1}{4} = 2y - 5$

2. $4y - \frac{2y}{5} + \frac{2}{3} = \frac{7}{5} - 3y$

3. $\frac{5y}{8} - \frac{y}{3} - \frac{1}{9} = \frac{3y}{4} - 2$

4. $\frac{7}{15} - \frac{5y}{12} = \frac{y}{6} - 5 - 4y$

When you have finished the pretest, correct it yourself using the answers on page 135. Then check all the questions you got wrong on the list below:

1. ______ 3. ______

2. ______ 4. ______

If you made any mistakes (even if it was only *one* mistake), read the introduction and go through the program.

If you made no mistakes at all, skip Sec. 2-6, because you already know its material, and go on to Sec. 2-7.

Introduction

You will learn how to apply the methods of the first four sections to equations containing fractions but no parentheses. You begin by using the multiplication transformation with the lowest common denominator (LCD) to eliminate fractions.

FRAMES FOR SECTION 2-6

1 If equations contain fractions, first use a multiplication transformation to eliminate fractions and then solve as usual.

$$\frac{4y}{3} + 3 - \frac{y}{6} = \frac{1}{3} + \frac{y}{2}$$

$$\Big\updownarrow \text{Multiplying both sides by 6}$$

$$8y + 18 - y = 2 + 3y$$

which can be solved by the usual methods to give {______}. Check your answer.

$\{-4\}$

CHECK

$$\frac{4}{3}(-4) + 3 - \frac{-4}{6} = \frac{1}{3} + \frac{-4}{2}$$

$$-\frac{16}{3} + 3 + \frac{2}{3} = \frac{1}{3} - 2$$

$$-\frac{14}{3} + 3 = \frac{1}{3} - 2$$

$$-\frac{5}{3} = -\frac{5}{3} \qquad \text{True}$$

---

2 Use a multiplication transformation to eliminate fractions from the following equations:

a. $\frac{3y}{5} + \frac{4}{7} + \frac{y}{7} = 4y + \frac{11}{5}$

$$\Big\updownarrow \text{MT with 35}$$

____________ = ____________

b. $\frac{3y}{8} + 2 + \frac{y}{4} = 6y + \frac{7}{2}$

$$\Big\updownarrow \text{MT with \_\_\_\_\_\_}$$

____________ = ____________

---

a. $21y + 20 + 5y = 140y + 77$ *b.* 8; $3y + 16 + 2y = 48y + 28$

---

3 Use a multiplication transformation to eliminate fractions from the following equations:

a. $\frac{2y}{7} - \frac{y}{3} + \frac{4}{7} = 4y - 6$

$\updownarrow$ MT with ____

______ = ______

b. $\frac{4y}{5} - \frac{y}{4} + \frac{3}{2} = 2y - \frac{11}{5}$

$\updownarrow$ MT with ____

______ = ______

---

a. 21; $6y - 7y + 12 = 84y - 126$
b. 20; $16y - 5y + 30 = 40y - 44$

---

4 Solve

$$\frac{4y}{11} + 5 = \frac{2}{5} + 4y$$

and check your answer. Show all your work.

$$\frac{4y}{11} + 5 = \frac{2}{5} + 4y$$

$\updownarrow$ MT with 55

$$20y + 275 = 22 + 220y$$

$$275 - 22 = 220y - 20y$$

$$253 = 200y$$

$$\frac{253}{200} = y$$

$$\left\{\frac{253}{200}\right\}$$

CHECK

$$\frac{\overset{1}{\cancel{4}}}{\underset{1}{\cancel{11}}}\left(\frac{\overset{23}{\cancel{253}}}{\underset{50}{\cancel{200}}}\right) + 5 = \frac{2}{5} + \overset{1}{\cancel{4}}\left(\frac{253}{\underset{50}{\cancel{200}}}\right)$$

$$\frac{23}{50} + 5 = \frac{2}{5} + \frac{253}{50}$$

$$\frac{23}{50} + \frac{250}{50} = \frac{20}{50} + \frac{253}{50}$$

$$\frac{273}{50} = \frac{273}{50} \qquad \text{True}$$

5 Solve

$$\frac{2k}{3} + 7 = \frac{2}{3} - \frac{11k}{12}$$

and check your answer. Show all your work in full.

$$\frac{2k}{3} + 7 = \frac{2}{3} - \frac{11k}{12}$$

$$\Updownarrow \text{MT with 12}$$

$$8k + 84 = 8 - 11k$$
$$8k + 11k = 8 - 84$$
$$19k = -76$$
$$k = -4$$
$$\{-4\}$$

CHECK

$$\frac{2}{3}(-4) + 7 = \frac{2}{3} - \frac{11}{12}(-4)$$

$$-\frac{8}{3} + 7 = \frac{2}{3} + \frac{11}{3}$$

$$-\frac{8}{3} + \frac{21}{3} = \frac{13}{3}$$

6 What is the lowest common denominator of $\frac{1}{15}$, $\frac{3}{5}$, and $-\frac{7}{9}$, that is, the lowest number into which 15, 5, and 9 will all divide?

45

7 What is the lowest common denominator of $\frac{3}{16}$, $-\frac{5}{12}$, and $\frac{1}{9}$?

144

---

8 In order to eliminate fractions, the number with which you multiply both sides of the equation must be a common denominator; and you *must* check before carrying out the multiplication that every denominator does in fact divide into your common denominator. It is also worth making some effort to find the *lowest* common denominator (LCD) in order to keep the numbers as small as possible. The LCD of $\frac{5}{12}$, $\frac{1}{4}$, $\frac{1}{9}$, and $-\frac{7}{9}$ is ______. Solve

$$\frac{5y}{12} + \frac{y}{4} - 6 = \frac{y}{9} - 5 - \frac{7}{9}$$

and *check your answer.*

36

$$\frac{5y}{12} + \frac{y}{4} - 6 = \frac{y}{9} - 5 - \frac{7}{9}$$

$$\updownarrow \text{MT with 36}$$

$$15y + 9y - 216 = 4y - 180 - 28$$
$$15y + 9y + (-4y) = -180 - 28 + 216$$
$$20y = 8$$
$$y = \frac{8}{20} = \frac{2}{5}$$

CHECK

$$\frac{5}{12}\left(\frac{2}{5}\right) + \frac{1}{4}\left(\frac{2}{5}\right) - 6 = \frac{1}{9}\left(\frac{2}{5}\right) - 5 - \frac{7}{9}$$
$$\frac{1}{6} + \frac{1}{10} - 6 = \frac{2}{45} - 5 - \frac{7}{9}$$
$$\frac{5 + 3 - 180}{30} = \frac{2 - 225 - 35}{45}$$
$$-\frac{172}{30} = -\frac{258}{45}$$
$$-\frac{86}{15} = -\frac{86}{15} \quad \text{True}$$

---

9 What is the LCD of $\frac{4}{7}, \frac{1}{2}, \frac{9}{14}$, and $-\frac{3}{4}$? Solve

$$\frac{4k}{7} + \frac{k}{2} - 3 = \frac{9}{14} - \frac{3k}{4}$$

and check your answer.

LCD is 28

MT with 28 gives

$$\begin{aligned} 16k + 14k - 84 &= 18 - 21k \\ 16k + 14k + 21k &= 18 + 84 \\ 51k &= 102 \\ k &= 2 \end{aligned}$$

$$\{2\}$$

CHECK

$$\begin{aligned} \frac{4}{7}(2) + \frac{2}{2} - 3 &= \frac{9}{14} - \frac{3}{4}(2) \\ \frac{8}{7} + 1 - 3 &= \frac{9}{14} - \frac{3}{2} \\ \frac{8 + 7 - 21}{7} &= \frac{9 - 21}{14} \\ -\frac{6}{7} &= -\frac{12}{14} \qquad \text{True} \end{aligned}$$

---

10 Solve

$$\frac{3z}{5} + \frac{5z}{6} - 5 = \frac{2}{15} + \frac{z}{3}$$

and check your answer.

MT with 30 gives

$$\begin{aligned} 18z + 25z - 150 &= 4 + 10z \\ 18z + 25z + (-10z) &= 4 + 150 \\ 33z &= 154 \\ z &= \frac{\overset{14}{\cancel{154}}}{\underset{3}{\cancel{33}}} = \frac{14}{3} \end{aligned}$$

$$\left\{\frac{14}{3}\right\}$$

CHECK

$$\frac{\overset{1}{\cancel{3}}}{5}\left(\frac{14}{\underset{1}{\cancel{3}}}\right) + \frac{5}{\underset{3}{\cancel{6}}}\left(\frac{\overset{7}{\cancel{14}}}{3}\right) - 5 = \frac{2}{15} + \frac{1}{3}\left(\frac{14}{3}\right)$$

$$\frac{14}{5} + \frac{35}{9} - 5 = \frac{2}{15} + \frac{14}{9}$$

$$\frac{126 + 175 - 225}{45} = \frac{6 + 70}{45}$$

$$\frac{76}{45} = \frac{76}{45} \qquad \text{True}$$

11 Solve

$$\frac{5b}{3} + \frac{3}{7} - \frac{b}{2} = \frac{3b}{14} + \frac{4}{21}$$

and check your answer.

MT with 42 gives

$$70b + 18 - 21b = 9b + 8$$

$$70b - 21b + (-9b) = 8 + (-18)$$

$$40b = -10$$

$$b = -\frac{1}{4}$$

$$\left\{-\frac{1}{4}\right\}$$

CHECK

$$\frac{5}{3}\left(-\frac{1}{4}\right) + \frac{3}{7} - \frac{1}{2}\left(-\frac{1}{4}\right) = \frac{3}{14}\left(-\frac{1}{4}\right) + \frac{4}{21}$$

$$-\frac{5}{12} + \frac{3}{7} + \frac{1}{8} = -\frac{3}{56} + \frac{4}{21}$$

$$\frac{-70 + 72 + 21}{168} = \frac{-9 + 32}{168}$$

$$\frac{23}{168} = \frac{23}{168} \qquad \text{True}$$

12 Solve

$$\frac{e}{6} + \frac{7}{8} - \frac{4e}{5} = 1 - \frac{7e}{12}$$

and check your answer.

MT with 120 gives

$$20e + 105 - 96e = 120 - 70e$$
$$20e - 96e + 70e = 120 + (-105)$$
$$-6e = 15$$
$$e = -\frac{15}{6} = -\frac{5}{2}$$
$$\left\{-\frac{5}{2}\right\}$$

CHECK

$$\frac{1}{6}\left(-\frac{5}{2}\right) + \frac{7}{8} - \frac{4}{5}\left(-\frac{5}{2}\right) = 1 - \frac{7}{12}\left(-\frac{5}{2}\right)$$
$$-\frac{5}{12} + \frac{7}{8} + 2 = 1 + \frac{35}{24}$$
$$\frac{-10 + 21 + 48}{24} = \frac{24 + 35}{24}$$
$$\frac{59}{24} = \frac{59}{24} \quad \text{True}$$

You have finished the program for Sec. 2-6. Now take the posttest to see how much you have learned.

POSTTEST 2-6

Solve the following questions and check your answers:

1. $\frac{5y}{6} - \frac{7y}{12} + \frac{1}{4} = 4y - 1$

2. $y - \frac{11y}{7} + \frac{7}{3} = \frac{3}{7} - 2y$

3. $\frac{3y}{5} - \frac{y}{8} - \frac{7}{4} = \frac{7y}{10} - 3$

4. $\frac{7}{16} - \frac{11y}{12} = \frac{y}{8} - 3 - 2y$

When you have finished the posttest, correct it yourself using the answers on page 135. Then check all the questions you got wrong on the list below:

1. ______ 3. ______

2. ______ 4. ______

If you compare your performance with that on Pretest 2-6, you will see how much you have improved. If you made more than one mistake on this posttest, review the material and take the pretest again. Don't move on to the next section until you can get all the questions right.

Now proceed to Sec. 2-7.

Section 2-7
Solving Equations Containing Parentheses and Fractions

Do this section if you made a mistake in Questions 7 to 9 of Diagnostic Test 2. Begin by taking the pretest below, which will show you how much of Sec. 2-7 you already know and what you will be able to do by the time you have finished it.

PRETEST 2-7

Solve the following equations and check your answers:

1. $\frac{1}{4}(4y + 3) + \frac{2}{3} = \frac{1}{6}(3y - 2)$

2. $\frac{1}{5}(10y + 7) - \frac{2}{3}(4y - 2) = \frac{1}{10}(5y + 4)$

3. $\frac{1}{2}\left(3y + \frac{4}{3}\right) - \frac{2}{3}\left(\frac{5y}{4} + 2\right) = 2y$

4. $\frac{1}{8}(2y + 3) - \frac{2}{3}\left(5y - \frac{1}{2}\right) = \frac{1}{4}\left(\frac{5}{6} - 3y\right)$

When you have finished the pretest, correct it yourself using the answers on page 135. Then check all the questions you got wrong on the list below:

1. ______ 3. ______

2. ______ 4. ______

If you made any mistakes (even if it was only *one* mistake), read the introduction and go through the program.

If you made no mistakes at all, skip Sec. 2-7, because you already know its material, and go on to the next section whose number you circled on page 62. If no further numbers were circled, take Final Test 2 (page 132).

Introduction

Now you can use all the methods of the earlier sections to solve equations with both fractions and parentheses. You will do best to adopt a regular procedure; and this section uses that of removing parentheses first and fractions second.

FRAMES FOR SECTION 2-7

1 With equations which contain parentheses and fractions it is best to proceed as shown:

$$\frac{3}{4}\left(3y+\frac{1}{2}\right)=\frac{5}{8}(2y+5)$$

↕ Remove parentheses by using the distributive law

$$\frac{9y}{4}+\frac{3}{8}=\frac{5y}{4}+\frac{25}{8}$$

↕ Remove fractions by using the MT with 8; that is, multiply both sides by 8

__________ = __________

which can be solved to give {________}.

$18y + 3 = 10y + 25$ $\left\{\frac{11}{4}\right\}$

---

2 $\frac{1}{9}(3y+5)+\frac{2}{3}=\frac{1}{2}(3y+4)$

↕ Remove parentheses by using the distributive law

__________ = __________

↕ Remove fractions by using the MT with 18

__________ = __________

↕ Solve by the usual method

$y =$ ________

Solution set is ________.

$$\frac{y}{3} + \frac{5}{9} + \frac{2}{3} = \frac{3y}{2} + 2$$

$$6y + 10 + 12 = 27y + 36$$

$$6y - 27y = 36 - 10 - 12$$

$$-21y = 14$$

$$y = -\frac{2}{3}$$

$$\left\{-\frac{2}{3}\right\}$$

---

3 $\frac{1}{5}(2y + 3) - \frac{1}{4}(y + 2) = \frac{3}{10}$

↕ Remove parentheses by using the distributive law

________________ = ________________

↕ Remove fractions by using the MT

________________ = ________________

↕ Solve by the usual method

y = _____

Solution set is ______.

$$\frac{2y}{5} + \frac{3}{5} - \frac{y}{4} - \frac{1}{2} = \frac{3}{10}$$

$$8y + 12 - 5y - 10 = 6$$

$$8y - 5y = 6 + 10 - 12$$

$$3y = 4$$

$$y = \frac{4}{3}$$

$$\left\{\frac{4}{3}\right\}$$

---

4 In checking, always go back to the original equation and work out the expressions within parentheses before removing the parentheses. Thus checking our answer to Frame 1 by substituting $\frac{11}{4}$ in

$$\frac{3}{4}\left(3y + \frac{1}{2}\right) = \frac{5}{8}(2y + 5)$$

we get

$$\frac{3}{4}\left(\frac{33}{4} + \frac{1}{2}\right) = \frac{5}{8}\left(\_\_ + \_\_\right)$$

$$\frac{3}{4}\left(\frac{33 + \_\_}{4}\right) = \frac{5}{8}\left(\frac{\_\_ + \_\_}{4}\right)$$

$$\frac{3}{4}\left(\frac{\_\_}{4}\right) = \frac{5}{8}\left(\frac{\_\_}{4}\right)$$

$$\frac{\_\_}{16} = \frac{\_\_}{32} \qquad \text{True}$$

$$\left(\frac{22}{4} + 5\right)$$

$$\frac{3}{4}\left(\frac{33 + 2}{4}\right) = \frac{5}{8}\left(\frac{22 + 20}{4}\right)$$

$$\frac{3}{4}\left(\frac{35}{4}\right) = \frac{5}{8}\left(\frac{42}{4}\right)$$

$$\frac{105}{16} = \frac{210}{32}$$

---

5 Check the answer to Frame 2 by substituting $-\frac{2}{3}$ in

$$\frac{1}{9}(3y + 5) + \frac{2}{3} = \frac{1}{2}(3y + 4)$$

This gives

$$\frac{1}{9}(\_\_\_\_\_) + \frac{2}{3} = \frac{1}{2}(\_\_\_\_\_)$$

$$\frac{\_}{9} + \frac{2}{3} = \frac{\_}{2}$$

$$\frac{\_ + \_}{3} = \_\_\_\_\_ \qquad \text{True}$$

$$\frac{1}{9}(-2 + 5) + \frac{2}{3} = \frac{1}{2}(-2 + 4)$$

$$\frac{3}{9} + \frac{2}{3} = \frac{2}{2}$$

$$\frac{1 + 2}{3} = 1 \qquad \text{True}$$

---

6 Check the answer to Frame 3 by substituting $\frac{4}{3}$ in

$$\frac{1}{5}(2y + 3) - \frac{1}{4}(y + 2) = \frac{3}{10}$$

Work out the expressions within parentheses first.

$$\frac{1}{5}\left(\frac{8}{3} + 3\right) - \frac{1}{4}\left(\frac{4}{3} + 2\right) = \frac{3}{10}$$

$$\frac{1}{5}\left(\frac{8 + 9}{3}\right) - \frac{1}{4}\left(\frac{4 + 6}{3}\right) = \frac{3}{10}$$

$$\frac{1}{5}\left(\frac{17}{3}\right) - \frac{1}{4}\left(\frac{10}{3}\right) = \frac{3}{10}$$

$$\frac{17}{15} - \frac{5}{6} = \frac{3}{10}$$

$$\frac{34 - 25}{30} = \frac{3}{10}$$

$$\frac{9}{30} = \frac{3}{10} \qquad \text{True}$$

---

7 $\frac{1}{7}(2y - 4) - \frac{1}{2}(y - 5) = \frac{3}{14}$

↕ Remove parentheses by using the distributive law

_______ = _______

↕ Remove fractions by using the MT

_______ = _______

↕ Solve

$y =$ ____

Solution set is ____.

$$\frac{2y}{7} - \frac{4}{7} - \frac{y}{2} + \frac{5}{2} = \frac{3}{14}$$

$$4y - 8 - 7y + 35 = 3$$

$$4y - 7y = 3 + 8 - 35$$

$$-3y = -24$$

$$y = 8$$

$$\{8\}$$

---

8 Check the answer to Frame 7 by substituting in the original equation. Work out the expressions within parentheses first.

$$\frac{1}{7}(16 - 4) - \frac{1}{2}(8 - 5) = \frac{3}{14}$$

$$\frac{12}{7} - \frac{3}{2} = \frac{3}{14}$$

$$\frac{24 - 21}{14} = \frac{3}{14} \quad \text{True}$$

---

9 Solve

$$\frac{1}{6}(5y + 4) - \frac{1}{3}(2y - 1) = \frac{1}{2}(3y + 4)$$

and check your answer.

$$\frac{5y}{6} + \frac{2}{3} - \frac{2y}{3} + \frac{1}{3} = \frac{3y}{2} + 2$$

$$5y + 4 - 4y + 2 = 9y + 12$$

$$5y - 4y - 9y = 12 - 4 - 2$$

$$-8y = 6$$

$$y = -\frac{3}{4}$$

$$\left\{-\frac{3}{4}\right\}$$

CHECK

$$\frac{1}{6}\left(-\frac{15}{4} + 4\right) - \frac{1}{3}\left(-\frac{3}{2} - 1\right) = \frac{1}{2}\left(-\frac{9}{4} + 4\right)$$

$$\frac{1}{6}\left(\frac{-15 + 16}{4}\right) - \frac{1}{3}\left(\frac{-3 - 2}{2}\right) = \frac{1}{2}\left(\frac{-9 + 16}{4}\right)$$

$$\frac{1}{6}\left(\frac{1}{4}\right) - \frac{1}{3}\left(-\frac{5}{2}\right) = \frac{1}{2}\left(\frac{7}{4}\right)$$

$$\frac{1}{24} + \frac{5}{6} = \frac{7}{8}$$

$$\frac{1 + 20}{24} = \frac{21}{24} \qquad \text{True}$$

---

10 Solve

$$\frac{1}{5}\left(\frac{3y}{2} + 10\right) = \frac{1}{4}\left(\frac{7y}{5} + 6\right)$$

and check your answer.

$$\frac{3y}{10} + 2 = \frac{7y}{20} + \frac{3}{2}$$

$$6y + 40 = 7y + 30$$

$$y = 10$$

$$\{10\}$$

CHECK

$$\frac{1}{5}(15 + 10) = \frac{1}{4}(14 + 6)$$

$$\frac{25}{5} = \frac{20}{4} \qquad \text{True}$$

---

11 Solve

$$\frac{3}{7}\left(\frac{2y}{9} + 5\right) = \frac{1}{3}\left(2y - \frac{5}{7}\right) + \frac{10}{3}$$

and check your answer.

$$\frac{2y}{21} + \frac{15}{7} = \frac{2y}{3} - \frac{5}{21} + \frac{10}{3}$$

$\updownarrow$ MT with 21

$$2y + 45 = 14y - 5 + 70$$

$$2y + (-14y) = -5 + 70 + (-45)$$

$$-12y = 20$$

$$y = -\frac{20}{12} = -\frac{5}{3}$$

$$\left\{-\frac{5}{3}\right\}$$

CHECK

$$\frac{3}{7}\left(-\frac{10}{27}+5\right)=\frac{1}{3}\left(-\frac{10}{3}-\frac{5}{7}\right)+\frac{10}{3}$$

$$\frac{\overset{1}{\cancel{3}}}{7}\left(\frac{-10+135}{\underset{9}{\cancel{27}}}\right)=\frac{1}{3}\left(\frac{-70-15}{21}\right)+\frac{10}{3}$$

$$\frac{125}{63}=-\frac{85}{63}+\frac{210}{63}\qquad\text{True}$$

12 Solve

$$\frac{5}{6}\left(2y-\frac{3}{4}\right)-\frac{2}{3}\left(\frac{5y}{4}-2\right)=\frac{1}{4}\left(3y+\frac{7}{3}\right)$$

and check your answer.

$$\frac{5y}{3}-\frac{5}{8}-\frac{5y}{6}+\frac{4}{3}=\frac{3y}{4}+\frac{7}{12}$$

$$\Big\updownarrow \text{ MT with 24}$$

$$40y-15-20y+32=18y+14$$

$$40y-20y+(-18y)=14+(-32)+15$$

$$2y=-3$$

$$y=-\frac{3}{2}$$

$$\left\{-\frac{3}{2}\right\}$$

CHECK

$$\frac{5}{6}\left(-3-\frac{3}{4}\right)-\frac{2}{3}\left(-\frac{15}{8}-2\right)=\frac{1}{4}\left(-\frac{9}{2}+\frac{7}{3}\right)$$

$$\frac{5}{6}\left(\frac{-12-3}{4}\right)-\frac{2}{3}\left(\frac{-15-16}{8}\right)=\frac{1}{4}\left(\frac{-27+14}{6}\right)$$

$$\frac{5}{6}\left(-\frac{15}{4}\right)-\frac{2}{3}\left(-\frac{31}{8}\right)=\frac{1}{4}\left(-\frac{13}{6}\right)$$

$$-\frac{75}{24}+\frac{62}{24}=-\frac{13}{24}\qquad\text{True}$$

You have finished the program for Sec. 2-7. Now take the posttest to see how much you have learned.

POSTTEST 2-7

Solve the following equations and check your answers:

1. $\frac{1}{3}(3y + 2) + \frac{3}{5} = \frac{1}{15}(5y - 6)$

2. $\frac{1}{4}(8y + 3) - \frac{1}{2}(4y - 3) = \frac{1}{6}(3y + 4)$

3. $\frac{1}{4}\left(\frac{5y}{3} + 2\right) - \frac{1}{3}\left(3y - \frac{5}{2}\right) = \frac{1}{6}(4y - 7)$

4. $\frac{3}{5}\left(6y + \frac{5}{3}\right) - \frac{2}{3}\left(\frac{4y}{5} + \frac{3}{2}\right) = \frac{4}{15}(2y - 7)$

When you have finished the posttest, correct it yourself using the answers on page 135. Then check all the questions you got wrong on the list below:

1. ______ 3. ______

2. ______ 4. ______

If you compare your performance with that on Pretest 2-7, you will see how much you have improved. If you made more than one mistake on this posttest, review the material and take the pretest again. Don't move on to the next section until you can get all the questions right.

Now proceed to the next section whose number you circled on page 62. If no other sections were circled, take Final Test 2 (page 132).

Section 2-8
Equations Which Are Closed Sentences

Do this section if you made a mistake in Questions 10 to 12 of the Diagnostic Test 2. Begin by taking the pretest below, which will show you how much of Sec. 2-8 you already know and what you will be able to do by the time you have finished it.

PRETEST 2-8

Give the solution sets of the following equations:

1. $3y + 8 - 5y - 3 = 13 - 2y - 8$
2. $6y + 11 + 3y - 2 = 6 + 5y + 7 + y - 4$
3. $2(4y + 5) = 7(2y + 5) + 3(11 - 2y)$
4. $6(2y - 5) - 5(2 - 3y) = 8(3y - 5)$
5. $\frac{2}{3}(y - 3) - \frac{3}{4}(3y - 5) = -\frac{1}{12}(y - 21)$
6. $\frac{2}{5}(3y + 8) - \frac{1}{3}(2y - 3) = \frac{4}{15}(2y + 3)$

When you have finished the pretest, correct it yourself using the answers on page 135. Then check all the questions you got wrong on the list below:

1. ______ 3. ______ 5. ______

2. ______ 4. ______ 6. ______

If you made any mistakes (even if it was only *one* mistake), read the introduction and go through the program.

If you made no mistakes at all, skip Sec. 2-8, because you already know its material, and go on to take Final Test 2.

Introduction

Section 2-8 explains the three special cases which sometimes cause confusion: (1) when the solution set is $\{0\}$; (2) when the solution set is $\emptyset$ because all values of the variable yield false statements on substitution; and (3) when the solution set is the domain of the variable because all values of the variable yield true statements on substitution:

FRAMES FOR SECTION 2-8

1 Consider the equation $5y + 3 = 5y + 3$.

a. Is $5(1) + 3 = 5(1) + 3$ a true statement?
b. Is 1 a solution of $5y + 3 = 5y + 3$?
c. Is $5(2) + 3 = 5(2) + 3$ a true statement?
d. Is 2 a solution of $5y + 3 = 5y + 3$?

---

a. yes *b.* yes *c.* yes *d.* yes

---

2 *a.* Is 3 a solution of $5y + 3 = 5y + 3$?
b. Is 0 a solution of $5y + 3 = 5y + 3$?
c. Is there any value of y which is not a solution of $5y + 3 = 5y + 3$? If so, what value?

---

a. yes *b.* yes *c.* no

---

3 Thus the solution set of $5y + 3 = 5y + 3$ is the domain of y or $\{y \mid y \text{ is a real number}\}$. What is the solution set of $4y + 2 = 4y + 2$

---

the domain of y, or $\{y \mid y \text{ is a real number}\}$

---

4 Notice that $4y + 2 = 4y + 2$ is not an open sentence but a closed sentence because it is always true.

$$4y + 2 = 4y + 2 \longleftrightarrow 4y - 4y = 2 - 2 \longleftrightarrow 0 = 0$$

which is a (true/false) ______ statement. But

$$4y + 2 = 4y + 3 \longleftrightarrow 4y - 4y = \_\_\_\_\_ \longleftrightarrow 0 = \_\_\_\_\_$$

which is a (true/false) ______ statement.

---

true; $3 - 2$; 1; false

---

5 Is there any value of y for which $4y + 2 = 4y + 3$ is true? What is the solution set of $4y + 2 = 4y + 3$?

---

no; $\emptyset$

---

6 Find the solution sets of:

a. $2y + 3 = 2y + 3$
b. $2y + 3 = 2y + 5$

---

a. the domain of y *b.* $\emptyset$

---

7 $2y + 3 = 3y + 3 \longleftrightarrow 2y - 3y = 3 - 3 \longleftrightarrow$ ____ $= 0 \longleftrightarrow$ ____ $= 0$

What is the solution set of $2y + 3 = 3y + 3$? *Check your answer.*

$-y$ y $\{0\}$

CHECK

$$2(0) + 3 = 3(0) + 3$$
$$3 = 3 \quad \text{True}$$

---

8 Find the solution sets of:

a. $3y + 8 = 3y + 9$
b. $3y + 8 = 4y + 8$
c. $3y + 8 = 3y + 8$

---

a. 0 *b.* $\{0\}$ *c.* the domain of y

---

9 What is the solution set of $7y - 3y + 11 = 23 + 2y - 12$? *Check your answer.*

$$7y - 3y + (-2y) = 23 - 12 + (-11)$$
$$2y = 0$$
$$y = 0$$
$$\{0\}$$

CHECK

$$7(0) - 3(0) + 11 = 23 + 2(0) - 12$$
$$11 = 23 - 12 = 11 \quad \text{True}$$

---

10 What is the solution set of $\frac{1}{3}(3y + 2) - \frac{1}{2}(4y + 5) = \frac{1}{6}(7 - 6y)$?

$$y + \frac{2}{3} - 2y - \frac{5}{2} = \frac{7}{6} - y$$

$\updownarrow$ MT with 6

$$6y + 4 - 12y - 15 = 7 - 6y$$
$$6y - 12y + 6y = 7 + (-4) + 15$$
$$0 = 18$$

Solution set is $\emptyset$.

---

11 What is the solution set of $\frac{1}{2}(5y + 3) - \frac{1}{4}(3y - 2) = \frac{1}{4}(7y + 8)$?

$$\frac{5y}{2} + \frac{3}{2} - \frac{3y}{4} + \frac{1}{2} = \frac{7y}{4} + 2$$

$\updownarrow$ MT with 4

$$10y + 6 - 3y + 2 = 7y + 8$$
$$10y - 3y + (-7y) = 8 + (-6) + (-2)$$
$$0 = 0 \qquad \text{Always true}$$

Solution set is the domain of y, or $\{y \mid y \text{ is a real number}\}$.

---

12 What is the solution set of $4(3y + 4) - 3(2y - 3) = 5(5 - y)$? *Check your answer.*

$$12y + 16 - 6y + 9 = 25 - 5y$$
$$12y - 6y + 5y = 25 + (-16) + (-9)$$
$$y = 0$$
$$\{0\}$$

CHECK

$$4(0 + 4) - 3(0 - 3) = 5(5)$$
$$16 + 9 = 25 \qquad \text{True}$$

---

13 *a.* If an equation is equivalent to a true statement such as $2 = 2$, $0 = 0$, etc., then the equation will be true for (all/no) ______ values of the variable; therefore its solution set will be ($\emptyset$/ the domain of the variable) __________.

b. If an equation is equivalent to a false statement, such as $1 = 3$, $5 = 0$, etc., then the equation will be true for (all/no) ______ values of the variable; therefore its solution set will be ($\emptyset$/ the domain of the variable) __________.

---

a. all; the domain of the variable *b.* no; $\emptyset$

---

14 What is the solution set of

$$\frac{7}{3}(y - 1) - \frac{1}{5}(7y - 3) = \frac{2}{15}(7y - 13)$$

$$\frac{7y}{3} - \frac{7}{3} - \frac{7y}{5} + \frac{3}{5} = \frac{14y}{15} - \frac{26}{15}$$

$$\updownarrow \text{ MT with 15}$$

$$35y - 35 - 21y + 9 = 14y - 26$$
$$35y - 21y + (-14y) = -26 + 35 + (-9)$$
$$0 = 0$$

Solution set is the domain of y, or $\{y|y \text{ is a real number}\}$.

---

15 Find the solution set of

$$\frac{1}{4}(5y + 3) - \frac{1}{12}(2y + 1) = \frac{2}{3}(7y + 1)$$

Check your answer.

$$\frac{5y}{4} + \frac{3}{4} - \frac{y}{6} - \frac{1}{12} = \frac{14y}{3} + \frac{2}{3}$$

$$\updownarrow \text{ MT with 12}$$

$$15y + 9 - 2y - 1 = 56y + 8$$
$$15y - 2y + (-56y) = 8 + (-9) + 1$$
$$-43y = 0$$
$$y = 0$$
$$\{0\}$$

CHECK

$$\frac{1}{4}(0+3) - \frac{1}{12}(0+1) = \frac{2}{3}(0+1)$$

$$\frac{3}{4} - \frac{1}{12} = \frac{2}{3}$$

$$\frac{9-1}{12} = \frac{8}{12} \qquad \text{True}$$

---

16 Find the solution set of

$$\frac{1}{6}(5y-2) + \frac{1}{2}(y-7) = \frac{2}{3}(2y+5)$$

$$\frac{5y}{6} - \frac{1}{3} + \frac{-y}{2} - \frac{7}{2} = \frac{4y}{3} + \frac{10}{3}$$

$\updownarrow$ MT with 6

$$5y - 2 + 3y - 21 = 8y + 20$$
$$5y + 3y + (-8y) = 20 + 2 + 21$$
$$0 = 43$$

Solution set is $\emptyset$

You have finished the program for Sec. 2-8. Now take the posttest to see how much you have learned.

POSTTEST 2-8

Give the solution sets of the following equations:

1. $4y + 3 - 7y + 2 = 11 - 3y + 8$
2. $3y - 5 + 6y - 7 = 2 + 11y - 14 - 2y$
3. $2(7y + 3) = 5(6y + 4) - 7(3y + 2)$
4. $4(3 - 5y) - 6(2 - 7y) = 11(2y + 5)$
5. $\frac{5}{6}(2y + 9) - \frac{3}{4}(5y - 2) = \frac{3}{4}(7y + 12)$
6. $\frac{3}{5}(6y + 2) - \frac{2}{3}(4y + 3) = \frac{2}{15}(7y - 6)$

When you have finished the posttest, correct it yourself using the answers on page 135. Then check all the questions you got wrong on the list below:

1. ______ 3. ______ 5. ______

2. ______ 4. ______ 6. ______

If you compare your performance with that on Pretest 2-8, you will see how much you have improved. If you made more than one mistake on this posttest, review the material and take the pretest again. Now proceed to Final Test 2.

FINAL TEST 2

Give the solution sets of the following equations and check your answers.

1. $\frac{5}{4}k = -\frac{3}{8}$

2. $-\frac{2}{9}y = -\frac{11}{4}$

3. $3y + 7 + 5y = 2y - 12 + 1$

4. $5d - 11 + 3 = 3d + 9 - 1 - 7d$

5. $5(2y - 3) - 7(3y - 2) = -2(3y + 2)$

6. $5(3m - 4) - (m + 3) = 7(4 - 3m)$

7. $\frac{3y}{5} - \frac{1}{4} + \frac{7y}{10} = \frac{3y}{2} + \frac{3}{4}$

8. $\frac{2}{3}(7y + 3) - \frac{4}{7}(2y - 3) = \frac{18}{7}(y + 1)$

9. $\frac{2}{5}\left(y + \frac{3}{4}\right) - \frac{3}{4}\left(\frac{3y}{5} + 2\right) = -\frac{1}{5}(3y + 7)$

10. $2(7y - 3) + 3(4 - 5y) = 6(1 - y)$

11. $y + 5 - 9y = 3 - 4(2y - 5) - 18$

12. $\frac{3}{4}(2y + 7) - \frac{5}{6}(y - 3) = \frac{1}{3}(2y - 5)$

The answers for this test are on page 136. Check the questions you got wrong on the list below and compare your results with those on Diagnostic Test 2 to see how much you have improved.

1. ______	5. ______	7. ______	10. ______
2. ______	6. ______	8. ______	11. ______
3. ______		9. ______	12. ______
4. ______			

If you made more than one mistake, review the material and then take the diagnostic test again. Make sure that you can get all the questions right before you finish the chapter. Now proceed to Chap. 3.

ANSWERS TO TESTS FOR CHAPTER 2

DIAGNOSTIC TEST 2

1. $\{-\frac{15}{14}\}$
2. $\{\frac{25}{36}\}$
3. $\{\frac{1}{7}\}$
4. $\{1\}$
5. $\{\frac{5}{6}\}$
6. $\{\frac{16}{19}\}$
7. $\{4\}$
8. $\{-3\}$
9. $\{-\frac{1}{3}\}$
10. $\{0\}$
11. domain of the variable
12. $\emptyset$

PRETEST 2-1

1. $z = -16$
2. $z = 9$
3. $z = 2$
4. $\{4\}$
5. $\{19\}$
6. $\{-14\}$

POSTTEST 2-1

1. $z = -9$
2. $z = 11$
3. $z = 4$
4. $\{-3\}$
5. $\{13\}$
6. $\{-10\}$

PRETEST 2-2

1. $k = -\frac{1}{2}$
2. $k = -\frac{7}{3}$
3. $k = \frac{1}{3}$
4. $\{\frac{13}{16}\}$
5. $\{\frac{20}{3}\}$
6. $\{\frac{21}{2}\}$

POSTTEST 2-2

1. $p = -4$
2. $p = -\frac{55}{21}$
3. $p = \frac{1}{4}$
4. $\{-\frac{10}{27}\}$
5. $\{\frac{15}{4}\}$
6. $\{\frac{72}{5}\}$

PRETEST 2-3

1. $\{-\frac{2}{3}\}$
2. $\{-\frac{5}{4}\}$
3. $\{5\}$
4. $\{\frac{7}{5}\}$
5. $\{\frac{7}{5}\}$
6. $\{\frac{11}{8}\}$

POSTTEST 2-3

1. $\{\frac{3}{2}\}$
2. $\{-\frac{2}{3}\}$
3. $\{2\}$
4. $\{\frac{1}{2}\}$
5. $\{-3\}$
6. $\{\frac{8}{9}\}$

PRETEST 2-4

1. $\{11\}$
2. $\{1\}$
3. $\{-6\}$
4. $\{-17\}$
5. $\{\frac{13}{9}\}$
6. $\{-\frac{5}{6}\}$

POSTTEST 2-4

1. $\{-1\}$
2. $\{\frac{1}{13}\}$
3. $\{\frac{7}{2}\}$
4. $\{-\frac{17}{2}\}$
5. $\{2\}$
6. $\{-\frac{1}{3}\}$

PRETEST 2-5

1. $\{\frac{18}{11}\}$
2. $\{\frac{5}{22}\}$
3. $\{-\frac{22}{3}\}$
4. $\{-\frac{3}{7}\}$
5. $\{-\frac{21}{2}\}$
6. $\{\frac{33}{16}\}$

POSTTEST 2-5

1. $\{\frac{30}{17}\}$
2. $\{\frac{5}{13}\}$
3. $\{\frac{11}{4}\}$
4. $\{-15\}$
5. $\{-\frac{5}{2}\}$
6. $\{\frac{13}{3}\}$

PRETEST 2-6

1. $\{3\}$
2. $\{\frac{1}{9}\}$
3. $\{\frac{136}{33}\}$
4. $\{-\frac{328}{205}\}$

POSTTEST 2-6

1. $\{\frac{1}{3}\}$
2. $\{-\frac{4}{3}\}$
3. $\{\frac{50}{9}\}$
4. $\{-\frac{165}{46}\}$

PRETEST 2-7

1. $\{-\frac{7}{2}\}$
2. $\{2\}$
3. $\{-\frac{1}{2}\}$
4. $\{\frac{3}{14}\}$

POSTTEST 2-7

1. $\{-\frac{5}{2}\}$
2. $\{\frac{19}{6}\}$
3. $\{2\}$
4. $\{-\frac{14}{19}\}$

PRETEST 2-8

1. domain of the variable
2. $\{0\}$
3. $\emptyset$
4. $\{0\}$
5. $\{0\}$
6. $\emptyset$

POSTTEST 2-8

1. $\emptyset$
2. domain of the variable
3. $\{0\}$
4. $\emptyset$
5. $\{0\}$
6. domain of the variable

FINAL TEST 2

1. $\{-\frac{3}{10}\}$
2. $\{\frac{99}{8}\}$
3. $\{-3\}$
4. $\{\frac{16}{9}\}$
5. $\{\frac{3}{5}\}$
6. $\{\frac{51}{35}\}$
7. $\{-5\}$
8. $\{-\frac{6}{5}\}$
9. $\{-\frac{4}{11}\}$
10. $\{0\}$
11. domain of the variable
12. $\emptyset$

Chapter 3
Inequations of the First Degree in One Variable

This chapter will show you how to extend the methods developed in Chaps. 1 and 2 to solve first-degree inequations in one variable. Section 3-1 establishes the validity of addition transformations and partial validity of multiplication transformations, i.e., for positive numbers only, and applies them to the solution of inequations. Section 3-2 shows you how to graph inequations. Then Sec. 3-3 shows how the solution sets of inequations can be checked. It also discusses how to solve inequations which are closed sentences.

Chapter 3 requires knowledge of Chaps. 1 and 2. Throughout this chapter the domain of the variable is assumed to be the set of real numbers.

DIAGNOSTIC TEST 3

Give the solution sets of the following inequations:

1. $2x - 7 < 6x + 5$

2. $-3x \geqslant 7$

Graph the solution sets of the following inequations on the number line:

3. $6y - 4 \leqslant 3y - 5$

4. $3x + 9 > 5x + 4$

5. $2x - 9 \neq 4x + 3$

Give the solution sets of the following inequations and check your answers:

6. $3x - 7 \leqslant 7x + 4$

7. $4x + 3 > 4x - 7$

8. $5x - 2 < 5x - 6$

When you have finished the diagnostic test, correct it yourself using the answers on page 164. Then check all the questions you got wrong on the list below:

1. ______	3. ______	6. ______
2. ______	4. ______	7. ______
	5. ______	8. ______

If you made no mistakes at all, you need not do any work in this chapter and should go straight on to Chap. 4, Equations Involving Absolute Value, which begins on page 169. But if you made any mistakes (even if it was only *one* mistake), follow these instructions:

	section	page
If you made any mistakes in Question 1 or 2, circle Sec. 3-1.	3-1	139
If you made any mistakes in Questions 3 to 5, circle Sec. 3-2.	3-2	148
If you made any mistakes in Questions 6 to 8, circle Sec. 3-3.	3-3	154

You should now work through the sections whose numbers you have circled, taking them in numerical order. The sections begin on the pages indicated.

Section 3-1
Solving Inequations

Do this section if you made a mistake in Question 1 or 2 of Diagnostic Test 3. Begin by taking the pretest below, which will show you how much of Sec. 3-1 you already know and what you will be able to do by the time you have finished it.

PRETEST 3-1

Give the solution sets of the following inequations:

1. $5x + 3 \geqslant 4x + 5$
2. $7x < -3$
3. $-6x \geqslant 5$
4. $4x - 5 \leqslant 9x - 7$
5. $2x - 9 > 7x + 4$

When you have finished the pretest, correct it yourself using the answers on page 164. Then check all the questions you got wrong on the list below:

1. ______ 4. ______

2. ______ 5. ______

3. ______

If you made any mistakes (even if it was only *one* mistake), read the introduction and go through the program.

If you made no mistakes at all, skip Sec. 3-1, because you already know its material, and go on to the next section whose number you circled on page 138. If no further numbers were circled, take Final Test 3 (page 163).

Introduction

Section 3-1 shows that addition transformations are valid for inequations as well as for equations, but multiplication transformations are valid for inequations only when a positive number is used. Multiplication of an inequation by a negative number alters its solution set. You learn to avoid such multiplications by using the identity that if $a < b$, then $-a > -b$. You can use this identity together with the addition and multiplication transformations to solve inequations (except those containing the $\neq$ sign).

FRAMES FOR SECTION 3-1

1 In the last topic we showed that adding a number or a multiple of the variable to both sides of an equation results in an equivalent equation, and we called this operation the addition transformation (AT); thus

a.

$$x = 1 \quad \updownarrow \text{ AT with } 1 \quad x + 1 = \_\_\_\_$$

b.

$$x = 1 \quad \updownarrow \text{ AT with } -1 \quad x + (-1) = \_\_\_\_$$

c.

$$x = 1 \quad \updownarrow \text{ AT with } 2x \quad 3x = \_\_\_\_ + 1$$

---

a. 2 *b.* 0 *c.* $2x$

---

2 If we apply the addition transformation to the inequality $3 > 1$, we can obtain a set of inequalities whose members include:

a.

$$3 > 1 \quad \downarrow \text{ AT with } 1 \quad 4 > \_\_\_\_$$

b.

$$3 > 1 \quad \downarrow \text{ AT with } 5 \quad 8 > \_\_\_\_$$

c.

$$3 > 1 \quad \downarrow \text{ AT with } -4 \quad -1 > \_\_\_\_$$

Notice that we are not assuming equivalence and therefore use a single arrow.

---

a. 2 *b.* 6 *c.* -3

---

3 Which of the statements $3 > 1$, $4 > 2$, $8 > 6$, and $-1 > -3$ are true?

---

all of them

---

4 Will all the inequalities derived from $3 > 1$ by addition transformations be true?

---

yes

---

5 In general it can be shown that any addition transformation on an inequality or an inequation results in an equivalent inequality or inequation. Thus:

a.
$$5 > 2$$
$$\updownarrow \text{ AT with } 6$$
$$11 > \underline{\qquad}$$

b.
$$x + 5 > 7$$
$$\updownarrow \text{ AT with } -5$$
$$x > \underline{\qquad}$$

c.
$$2x + 3 > x + 2$$
$$\updownarrow \text{ AT with } -x$$
$$x + 3 > \underline{\qquad}$$

---

a. 8 *b.* 2 *c.* 2

---

6 This gives us a method for solving inequations:

$$3x + 5 > 2x + 3$$
$$\updownarrow \text{ AT with } -2x$$
$$x + 5 > \underline{\qquad}$$
$$\updownarrow \text{ AT with } -5$$
$$x > \underline{\qquad}$$

Solution set is $\{x \mid x \text{ is a real number and } x > \underline{\qquad}\}$.

---

3 -2 -2

---

7 Use the method of addition transformation to solve $3x + 3 \leqslant 2x + 5$.

$$3x + 3 \leqslant 2x + 5$$

$\updownarrow$ AT with $-2x$

$$x + 3 \leqslant 5$$

$\updownarrow$ AT with -3

$$x \leqslant 2$$

Solution set is $\{x \mid x \text{ is a real number and } x \leqslant 2\}$.

---

8 Use the method of addition transformation to solve $5x - 2 \geqslant 4x - 6$.

$$5x - 2 \geqslant 4x - 6$$

$\updownarrow$ AT with $-4x$

$$x - 2 \geqslant -6$$

$\updownarrow$ AT with 2

$$x \geqslant -4$$

Solution set is $\{x \mid x \text{ is a real number and } x \geqslant -4\}$.

---

9 Now let us examine the effect of multiplication transformations on inequalities.

a. $3 > 2$

$\downarrow$ MT with 2

$6 > 4$

b. $3 > 2$

$\downarrow$ MT with 6

$18 > 12$

c. $3 > 2$

$\downarrow$ MT with -1

$-3 > -2$

d. $3 > 2$

$\downarrow$ MT with -3

$-9 > -6$

Which of these inequalities are true statements?

---

a and *b* are true, but *c* and *d* are false.

---

10 Similarly

a. $3 > -2$

$\downarrow$ MT with 2

$6 > \_\_\_\_$

b. $3 > -2$

$\downarrow$ MT with $\frac{1}{2}$

$\frac{3}{2} > \_\_\_\_$

c. $3 > -2$

$\downarrow$ MT with $-\frac{1}{2}$

$-\frac{3}{2} > \_\_\_\_$

d. $3 > -2$

$\downarrow$ MT with -3

$-9 > \_\_\_\_$

e. $3 > -2$

$\downarrow$ MT with 0

$0 > \_\_\_\_$

Which of these inequalities are true statements?

---

a. -4; true *b.* -1; true *c.* 1; false *d.* 6; false
e. 0; false

---

11 *a.* What kinds of multiplication transformation will change the truth of an inequality?

b. What kinds of multiplication transformation will not change the truth of an inequality?

---

a. MTs with negative numbers or with zero
b. MTs with positive numbers

---

12 In general it can be shown that a multiplication transformation changes an inequation into another inequation and that the two inequations are equivalent when the MT is with a ______ number but are not equivalent when the MT is with a ______ number.

---

positive; negative

---

13 Using this method to solve $3x > 6$,

$$3x > 6$$

$$\updownarrow \text{ MT with } \tfrac{1}{3}$$

$$x > \_\_\_\_\_$$

Solution set is $\{x \mid x \text{ is a real number and } x > \_\_\_\_\_\}$.

---

2 2

---

14 Use the same method to solve $4x \leqslant 12$.

$$4x \leqslant 12$$

$$\updownarrow \text{ MT with } \tfrac{1}{4}$$

$$x \leqslant 3$$

Solution set is $\{x \mid x \text{ is a real number and } x \leqslant 3\}$.

---

15 Using both the AT and the MT

$$5x + 2 < 7x + 9$$

$$\updownarrow \text{ AT with } -5x \text{ and } -9$$

$$2 + (-9) < 7x + (-5x) \longleftrightarrow \_\_\_\_\_ < \_\_\_\_\_$$

$$\updownarrow \text{ MT with } \tfrac{1}{2}$$

$$\_\_\_\_\_ < x$$

Solution set is __________________________.

$-7 < 2x \longleftrightarrow -\frac{7}{2} < x$ $\{x \mid x \text{ is a real number and } x > -\frac{7}{2}\}$

---

16 Similarly $13x + 4 \geqslant 9x + 3$

$$\updownarrow \text{ AT with } -9x \text{ and } -4$$

$$13x + (-9x) \geqslant 3 + (-4) \longleftrightarrow \_\_\_\_\_ \geqslant \_\_\_\_\_$$

$$\updownarrow \text{ MT with } \tfrac{1}{4}$$

$$x \geqslant \_\_\_\_\_\_\_$$

Solution set is __________________________.

$4x \geqslant -1 \longleftrightarrow x \geqslant -\frac{1}{4}$ $\{x \mid x \text{ is a real number and } x \geqslant -\frac{1}{4}\}$

17 Use the same method to solve $7x + 3 \leqslant 3x - 5$.

$$\begin{aligned} 7x + 3 &\leqslant 3x - 5 \\ 7x + (-3x) &\leqslant -5 + (-3) \\ 4x &\leqslant -8 \\ x &\leqslant -2 \end{aligned}$$

$\{x \mid x \text{ is a real number and } x \leqslant -2\}$

18 If you had chosen to assemble the terms containing the variable on the other side of the inequation, you would have obtained

$$\begin{aligned} 7x + 3 &\leqslant 3x - 5 \\ 3 + 5 &\leqslant 3x + (-7x) \\ 8 &\leqslant -4x \\ 2 &\leqslant -x \end{aligned}$$

But since you cannot carry out an MT with a negative number, you have to continue your solution by another method, such as

$$\begin{aligned} 2 &\leqslant -x \\ &\updownarrow \text{ AT with } x \\ 2 + x &\leqslant \underline{\qquad} \\ &\updownarrow \text{ AT with } -2 \\ x &\leqslant \underline{\qquad} \end{aligned}$$

How does this solution compare with that of Frame 17?

0 $\quad -2 \quad$ It is the same.

19 Fill in $<$ or $>$ to make the following statements true:

a. $2 \underline{\qquad} 3$

b. $-2 \underline{\qquad} -3$

c. $8 \underline{\qquad} 5$

d. $-8 \underline{\qquad} -5$

a. $<$ *b.* $>$ *c.* $>$ *d.* $<$

20 In general

a. If $a < b$, then $-a$ ______ $-b$.

b. If $a > b$, then $-a$ ______ $-b$.

c. If $a \geqslant b$, then $-a$ ______ $-b$.

---

a. $>$ *b.* $<$ *c.* $\leqslant$

---

21 This can be easily proved because

$$a < b$$

$$\updownarrow \text{ AT with } -a$$

$$\underline{\qquad} < \underline{\qquad}$$

$$\updownarrow \text{ AT with } -b$$

$$\underline{\qquad} < \underline{\qquad} \longleftrightarrow -a \ \underline{\qquad}\ -b$$

$0 < b + (-a) \longleftrightarrow -b < -a \longleftrightarrow -a > -b$

---

22 Use this property to solve $-5x < 7$.

$-5x < 7 \longleftrightarrow 5x > -7 \longleftrightarrow x > -\dfrac{7}{5}$

$\left\{x \mid x \text{ is a real number and } x > -\dfrac{7}{5}\right\}$

---

23 Use the same property to solve $-4x \geqq -6$.

$-4x \geqslant -6 \longleftrightarrow 4x \leqslant 6 \longleftrightarrow x \leqslant \dfrac{3}{2}$

$\left\{x \mid x \text{ is a real number and } x \leqslant \dfrac{3}{2}\right\}$

---

24 Solve $8x - 9 \geqslant 5x - 13$.

$8x - 9 \geqslant 5x - 13 \longleftrightarrow 8x + (-5x) \geqslant -13 + 9$

$\longleftrightarrow 3x \geqslant -4 \longleftrightarrow x \geqslant -\dfrac{4}{3}$

Solution set is $\left\{x \mid x \text{ is a real number and } x \geqslant -\dfrac{4}{3}\right\}$

---

25 Solve $-4x \leqslant -9$.

$$-4x \leqslant -9 \longleftrightarrow 4x \geqslant 9 \longleftrightarrow x \geqslant \frac{9}{4}$$

Solution set is $\left\{x \mid x \text{ is a real number and } x \geqslant \frac{9}{4}\right\}$.

You have finished the program for Sec. 3-1. Now take the posttest to see how much you have learned.

POSTTEST 3-1

Give the solution set of the following inequations:

1. $6x + 2 \leqslant 5x + 3$
2. $3x > -8$
3. $-7x \leqslant 4$
4. $2x - 6 \geqslant 7x - 3$
5. $4x + 7 < 9x - 4$

When you have finished the posttest, correct it yourself using the answers on page 165. Then check all the questions you got wrong on the list below:

1. ______ 4. ______

2. ______ 5. ______

3. ______

If you compare your performance with that on Pretest 3-1, you will see how much you have improved. If you made more than one mistake on this posttest, review the material and take the pretest again. Don't move on to the next section until you can get all the questions right.

Now proceed to the next section whose number you circled on page 138. If no other sections were circled, take Final Test 3 (page 163).

Section 3-2
Graphing Inequations

Do this section if you made a mistake in Questions 3 to 5 of Diagnostic Test 3. Begin by taking the pretest below, which will show you how much of Sec. 3-2 you already know and what you will be able to do by the time you have finished it.

PRETEST 3-2

Graph the solution sets of the following inequations on the number line:

1. $y \neq 2$
2. $y > -1$
3. $y \leqslant -2$
4. $5x + 2 < 2x - 4$
5. $3x - 5 \geqslant 7x - 6$
6. $2x + 7 \neq 4x + 1$

When you have finished the pretest, correct it yourself using the answers on page 165. Then check all the questions you got wrong on the list below:

1. _____ 3. _____ 5. _____

2. _____ 4. _____ 6. _____

If you made any mistakes (even if it was only *one* mistake), read the introduction and go through the program.

If you made no mistakes at all, skip Sec. 3-2, because you already know its material, and go on to the next section whose number you circled on page 138. If no further numbers are circled, take Final Test 3 (page 163).

Introduction

You will learn a notation for graphing inequations of the first degree in one variable and how to apply it to the graphing of open sentences of the form $y < a$, $y \leqslant a$, $y \neq a$, $y > a$, and $y \geqslant a$. Then you are shown how more complex inequations can be graphed after solving them by the methods of Sec. 3-1. Finally you see how inequations containing the $\neq$ sign are graphed through solving the corresponding equations.

FRAMES FOR SECTION 3-2

1 The graph of $x \neq 2$ is

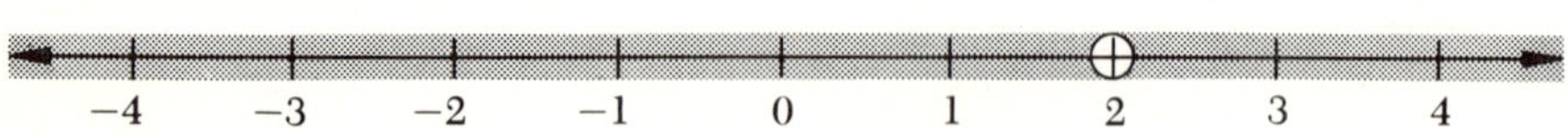

Draw the graph of $x \neq 3$.

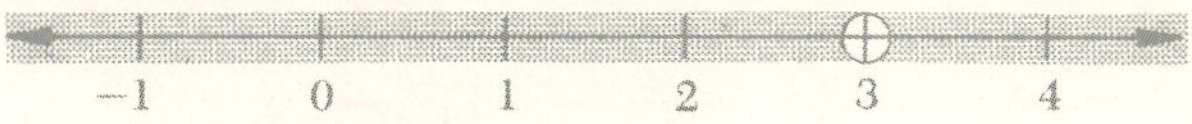

2 Draw the graph of $y \neq -1$.

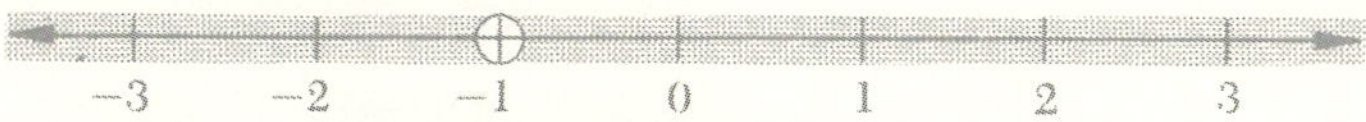

3 The graphs of $x < -2$ and $x > -2$ are shown below. Which graph corresponds to $x < -2$? Draw the graph of $x > 4$.

a.

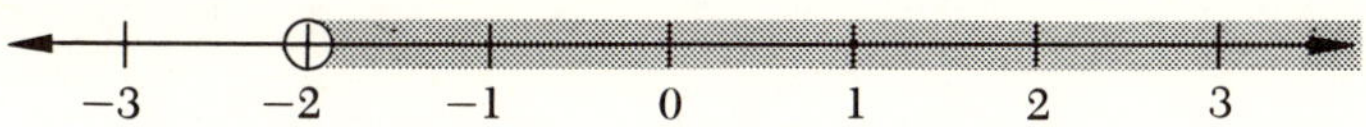

b.

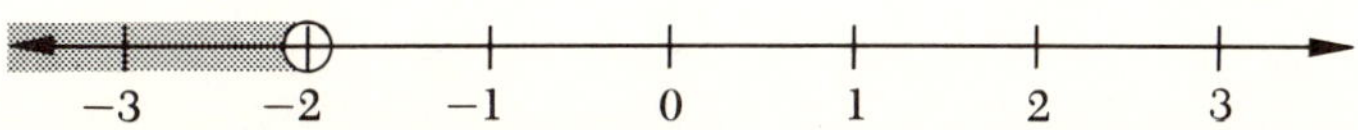

b

4 Draw the graph of $y < -3$.

5 The graphs of $y \leqslant 4$ and $y < 4$ are shown below. Which graph corresponds to $y \leqslant 4$? Draw the graph of $y \leqslant -2$.

a.

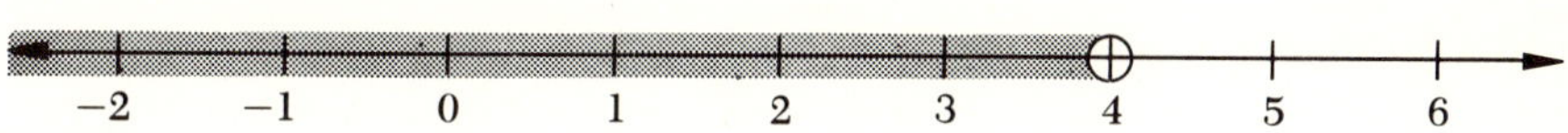

b.

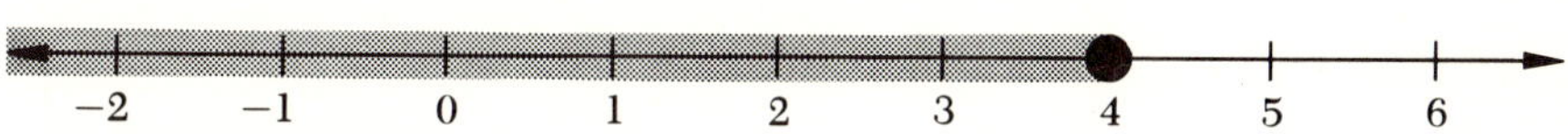

b

6 Draw the graph of $y \leqslant 0$.

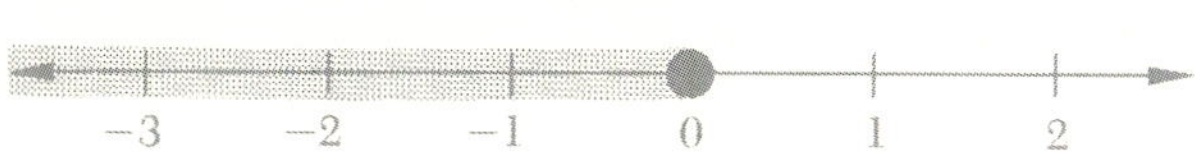

7 Which graph corresponds to $y \geqslant 3$? Draw the graph of $y \geqslant -1$.

a.
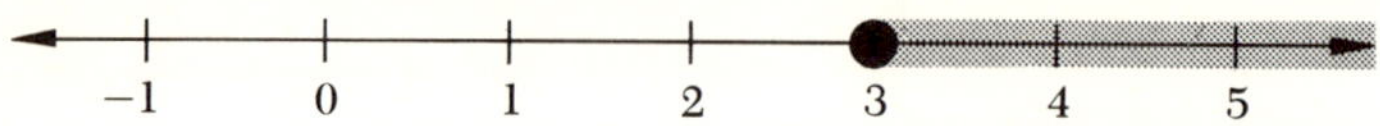

b.
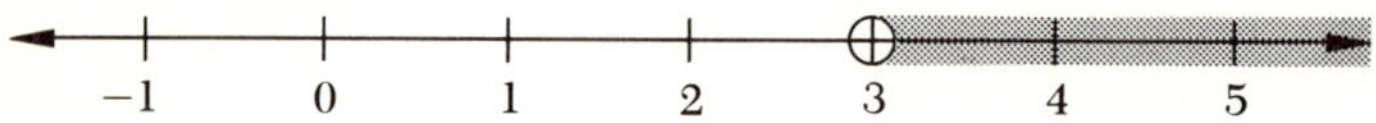

c.
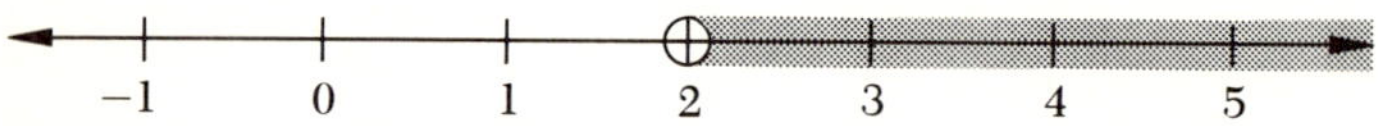

---

a

---

8 Draw the graphs of $y \leqslant 3$ and $y \geqslant -2$.

---

a.
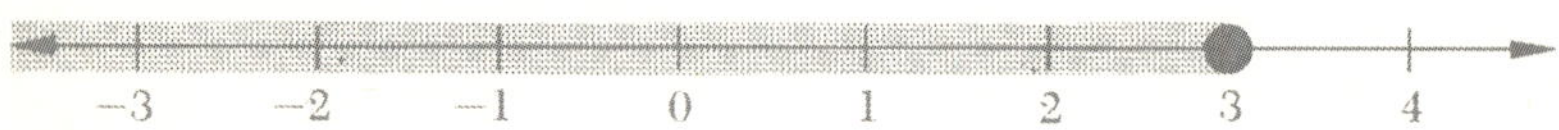

b.
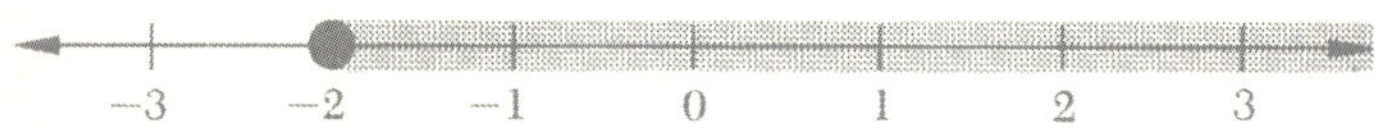

---

9 Draw the graphs of:

a. $y \neq 1$
b. $y \leqslant 2$
c. $y > -3$

---

a.
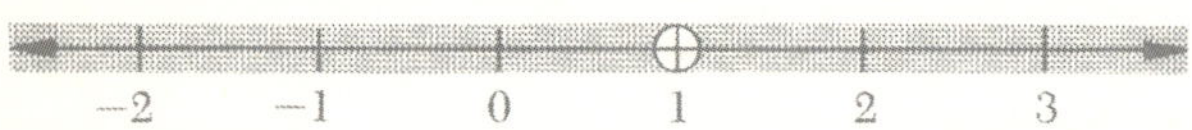

b.
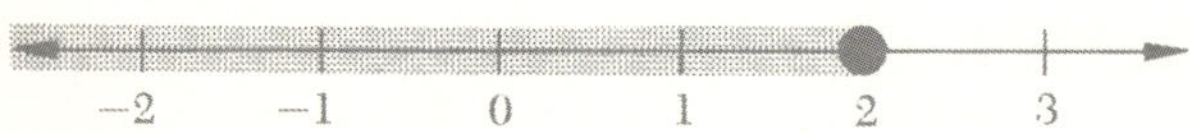

c.
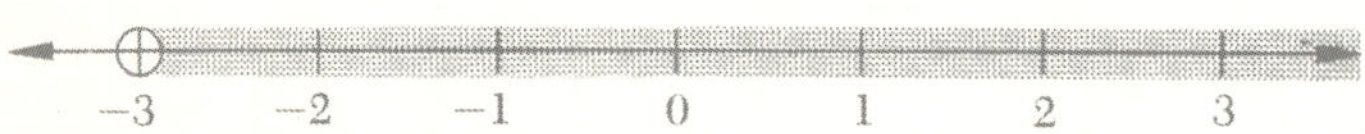

---

10 Solve the inequation $3x + 2 > 2x + 3$ and graph its solution set.

$3x + 2 > 2x + 3 \longleftrightarrow 3x + (-2x) > 3 + (-2) \longleftrightarrow x > 1$

$\{x \mid x \text{ is a real number and } x > 1\}$

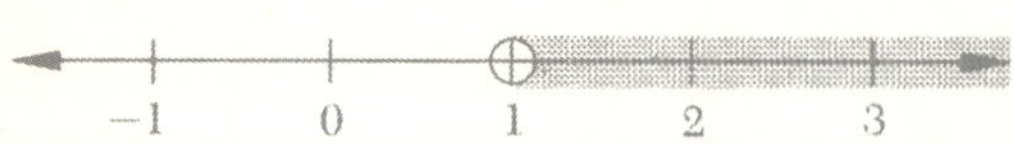

---

11 Solve the inequation $3x + 2 < 2x + 3$ and graph its solution set.

$\{x \mid x \text{ is a real number and } x < 1\}$

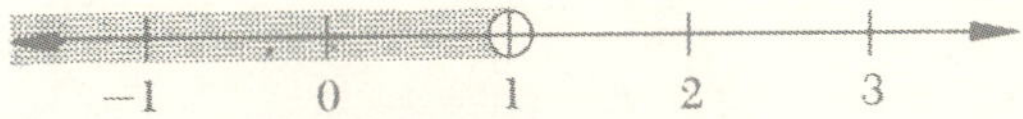

12 Solve the inequation $3x + 7 \geqslant 2x + 9$ and graph its solution set.

$\{x \mid x \text{ is a real number and } x \geqslant 2\}$

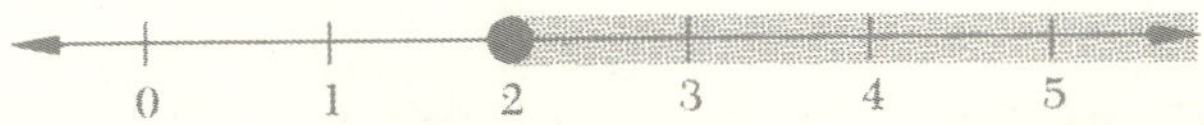

13 Solve the inequation $3x + 1 \leqslant 6x + 19$ and graph its solution set.

$\{x \mid x \text{ is a real number and } x \geqslant -6\}$

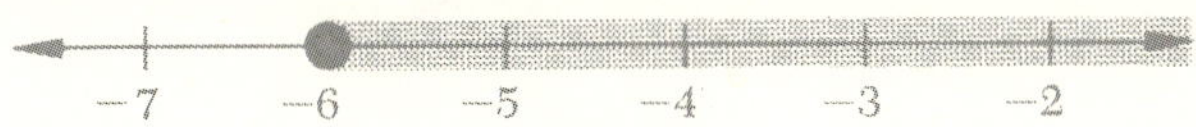

14 Solve the inequation $2x + 7 < 5x + 2$ and graph its solution set.

$\{x \mid x \text{ is a real number and } x > \frac{5}{3}\}$

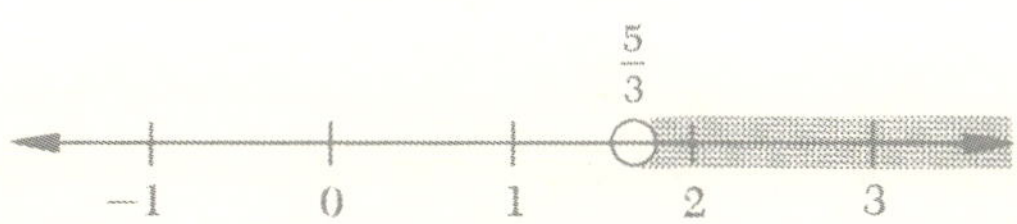

15 Solve the inequation $2x + 9 \geqslant 5x + 7$ and graph its solution set.

$\{x \mid x \text{ is a real number and } x \leqslant \frac{2}{3}\}$

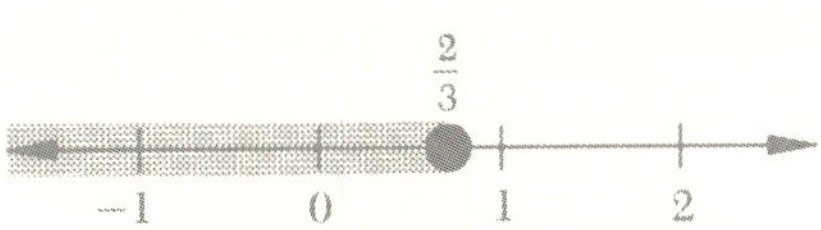

16 Solve the equation $2x - 5 = 5x - 2$ and graph its solution set.

$\{-1\}$

17 Graph the solution set of $2x - 5 \neq 5x - 2$.

---

18 Solve the inequation $5x - 9 \neq 7x + 4$ and graph its solution set. **Hint:** First solve the corresponding equation.

$\{x \mid x \text{ is a real number and } x \neq -\frac{13}{2}\}$

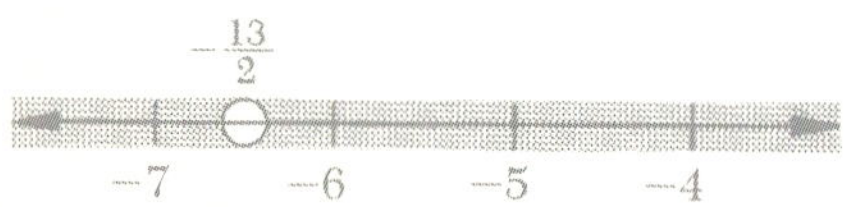

---

19 Solve the inequation $2x + 3 \neq 16x - 5$ and graph its solution set.

$\{x \mid x \text{ is a real number and } x \neq \frac{4}{7}\}$

---

20 Solve the inequation $x - 5 \neq 2 - 3x$ and graph its solution set.

$\{x \mid x \text{ is a real number and } x \neq \frac{7}{4}\}$

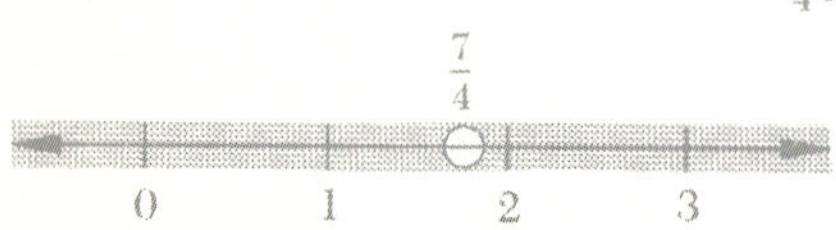

You have finished the program for Sec. 3-2. Now take the posttest to see how much you have learned.

POSTTEST 3-2

Graph the solution sets of the following inequations on the number line.

1. $y \neq 3$
2. $y > -4$
3. $y \leqslant -3$
4. $7x - 3 < 3x - 11$
5. $12x - 5 \geqslant 4x + 1$
6. $2x + 9 \neq 5x + 4$

When you have finished the posttest, correct it yourself using the answers on page 165. Then check all the questions you got wrong on the list below:

1. _____ 4. _____

2. _____ 5. _____

3. _____ 6. _____

If you compare your performance with that on Pretest 3-2, you will see how much you have improved. If you made more than one mistake on this posttest, review the material and take the pretest again. Don't move on to the next section until you can get all the questions right.

Now proceed to the next section whose number you circled on page 138. If no other sections were circled, take Final Test 3 (page 163).

Section 3-3
Checking Solutions and Special Cases

Do this section if you made a mistake in Questions 6 to 8 of Diagnostic Test 3. Begin by taking the pretest below, which will show you how much of Sec. 3-3 you already know and what you will be able to do by the time you have finished it.

PRETEST 3-3

Give the solution sets of the following inequations and check your answers:

1. $3x - 4 \leqslant 7x + 3$
2. $2x + 7 > 2x + 5$
3. $5x + 2 < 9x + 5$
4. $3x - 4 \geqslant 3x + 7$
5. $6x - 5 \geqslant 3x - 3$

When you have finished the pretest, correct it yourself using the answers on page 165. Then check all the questions you got wrong on the list below:

1. ______ 4. ______

2. ______ 5. ______

3. ______

If you made any mistakes (even if it was only *one* mistake), read the introduction and go through the program.

If you made no mistakes at all, skip Sec. 3-3, because you already know its material, and go on to take Final Test 3 (page 163).

Introduction

After the boundary value of an inequation that is an open sentence is defined, you are shown that it is a solution of the corresponding equation. Then you learn how the solution set of such an inequation can be checked by successive substitutions of the boundary value and one other value (usually 0). Solutions of closed sentences are then discussed, and you see how to check them by the substitution of 0.

FRAMES FOR SECTION 3-3

1 Solve the inequation $3x - 9 > 5x + 3$ and graph its solution set.

$$\begin{aligned} 3x - 9 &> 5x + 3 \\ -9 + (-3) &> 5x + (-3x) \\ -12 &> 2x \\ -6 &> x \end{aligned}$$

$\{x \mid x \text{ is a real number and } x < -6\}$

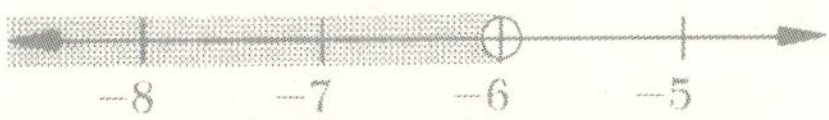

---

2 We call -6 the *boundary value* of $3x - 9 > 5x + 3$. What is the boundary value of $2x + 7 > 8x - 11$?

3, because its solution set is $\{x \mid x \text{ is a real number and } x < 3\}$.

---

3 If we substitute the boundary value -6 for x in each side of the inequation $3x - 9 > 5x + 3$, we get

	$3x - 9 = 3(-6) - 9 =$ ______
and	$5x + 3 = 5(-6) + 3 =$ ______
Therefore	$3x - 9 = 5x + 3$ when $x =$ ______

and the solution set of the equation $3x - 9 = 5x - 3$ is ______.

-27 -27 -6 $\{-6\}$

---

4 If we substitute the boundary value 3 for x in each side of the inequation $2x + 7 > 8x - 11$, we get

	$2x + 7 = 2(3) + 7 =$ ______
and	$8x - 11 = 8(3) - 11 =$ ______
Therefore	$2x + 7 = 8x - 11$ when $x =$ ______

and the solution set of the equation $2x + 7 = 8x - 11$ is ______.

---

13 13 3 $\{3\}$

---

5 In general the boundary value of an inequation of the first degree in one variable is the solution of the corresponding equation. The reason for this becomes apparent when we solve an inequation and its corresponding equation side by side.

$3x + 5 > 2x + 3$	$3x + 5 = 2x + 3$
$3x + (-2x) > 3 + (-5)$	$3x + (-2x) = 3 + (-5)$
$x >$ ______	$x =$ ______
Boundary value of inequation is ______.	Solution set of equation is ______.

$-2 \quad -2 \quad -2 \quad \{-2\}$

6 We can use this property to check the solution set of an inequation. Thus to check that $\{x \mid x \text{ is a real number and } x \leq \frac{16}{3}\}$ is the solution set of $2x + 7 \geq 5x - 9$, we substitute $\frac{16}{3}$ for x:

a. $2x + 7 = 2\left(\frac{16}{3}\right) + 7 =$ ______

b. $5x - 9 = 5\left(\frac{16}{3}\right) - 9 =$ ______

c. Is $\frac{16}{3}$ the correct boundary value?

a. $\frac{53}{3}$ *b.* $\frac{53}{3}$ *c.* yes

7 Check that $-\frac{6}{5}$ is the correct boundary value of the inequation $2x + 5 \leq 7x + 11$.

$$2x + 5 = 2\left(-\frac{6}{5}\right) + 5 = \frac{-12 + 25}{5} = \frac{13}{5}$$

$$7x + 11 = 7\left(-\frac{6}{5}\right) + 11 = \frac{-42 + 55}{5} = \frac{13}{5}$$

Therefore $-\frac{6}{5}$ is the correct boundary value.

8 Checking the boundary value is not a complete check, because in Frame 6 the set $A = \{x \mid x \text{ is a real number and } x \geq \frac{16}{3}\}$ also has a boundary value of $\frac{16}{3}$. To distinguish between set A and set $B = \{x \mid x \text{ is a real number and } x \leq \frac{16}{3}\}$, another substitution is necessary.

If, therefore, we substitute 0 for x in $2x + 7 \geq 5x - 9$, we get $0 + 7 \geq 0 - 9$ which is (true/false) ______.

Thus 0 (is/is not) ______ a solution of $2x + 7 \geq 5x - 9$, and the solution set (must/must not) ______ contain 0.

Therefore the solution set is (A/B) ______.

---

true; is; must; B

---

9 If 3 is the boundary value of $3x + 5 > 2x + 8$, we know that the solution set must either be $A = \{x \mid x \text{ is a real number and } x > 3\}$ or $B = \{x \mid x \text{ is a real number and } x < 3\}$.

Substituting 0 for x, gives ______ > ______, which is (true/false) ______.

Thus 0 (is/is not) ______ a solution and the solution set (must/must not) ______ contain 0. Therefore the solution set is ______ (A/B).

$5 > 8$, false; is not; must not; A

---

10 -3 is the boundary value of the inequation $5x + 8 < 2x - 1$. Use the method of substituting 0 for x to find whether the solution set is $A = \{x \mid x \text{ is a real number and } x > -3\}$ or $B = \{x \mid x \text{ is a real number and } x < -3\}$.

$0 + 8 < 0 - 1 \longleftrightarrow 8 < -1$ false

0 is not a solution, and the solution set must not contain 0. Therefore the solution set is B.

---

11 In Frame 7 you found that $-\dfrac{6}{5}$ was the correct boundary value of the inequation $2x + 5 \leq 7x + 11$. Now find the solution set by substituting 0 for x.

$0 + 5 \leq 0 + 11 \longleftrightarrow 5 \leq 11$ true

0 is a solution and the solution set must contain 0. Therefore the solution set is $\{x \mid x \text{ is a real number and } x \geq -\frac{6}{5}\}$.

---

12 Normally we use substitution for *checking* rather than finding the solution set. Thus to confirm our work in Frame 17 of Sec. 3-1, which found that the solution set of the inequation $7x + 3 \leqslant 3x - 5$ is $\{x \mid x \text{ is a real number and } x \leqslant -2\}$,

a. Substitute with -2, the boundary value

$$7x + 3 = 7(-2) + 3 = \underline{\qquad}$$
$$3x - 5 = \underline{\qquad} = \underline{\qquad}$$

b. Substitute with 0

$$7x + 3 = 0 + 3 = \underline{\qquad}$$
$$3x - 5 = \underline{\qquad}$$

c. Is 0 a solution of $7x + 3 \leqslant 3x - 5$?

d. Is 0 an element in $\{x \mid x \text{ is a real number and } x \leqslant -2\}$?

e. Is the proposed solution set confirmed?

---

a. $-11; 3(-2) - 5; -11$ *b.* $3; -5$ *c.* no *d.* no *e.* yes

---

13 Use the same method to confirm our work in Frame 16 of Sec. 3-1, which found that the solution set of the inequation $13x + 4 \geqslant 9x + 3$ is $\{x \mid x \text{ is a real number and } x \geqslant -\frac{1}{4}\}$.

Substituting with $-\dfrac{1}{4}$,

$$13\left(-\frac{1}{4}\right) + 4 = \frac{-13 + 16}{4} = \frac{3}{4}$$

$$9\left(-\frac{1}{4}\right) + 3 = \frac{-9 + 12}{4} = \frac{3}{4}$$

Substituting with 0 gives $4 \geqslant 3$, which is true. So 0 is a solution. 0 is also an element in $\{x \mid x \text{ is a real number and } x > -\frac{1}{4}\}$, and so the proposed solution set is confirmed.

---

14 Solve $5x - 3 > 2x - 7$ and check your answer.

$$5x - 3 > 2x - 7$$
$$5x + (-2x) > -7 + 3$$
$$3x > -4$$
$$x > -\frac{4}{3}$$

Solution set is $\{x \mid x \text{ is a real number and } x > -\frac{4}{3}\}$.

CHECK

$$5\left(-\frac{4}{3}\right) - 3 = \frac{-20 - 9}{3} = -\frac{29}{3} \qquad 2\left(-\frac{4}{3}\right) - 7 = \frac{-8 - 21}{3} = -\frac{29}{3}$$

Since $-3 > -7$ is true, 0 is a solution; $0 > -\frac{4}{3}$ and is an element in the solution set.

15 **Solve $-7x \leqslant 5$ and check your answer.**

$$-7x \leqslant 5$$
$$7x \geqslant -5$$
$$x \geqslant -\frac{5}{7}$$

Solution set is $\{x \mid x \text{ is a real number and } x \geqslant -\frac{5}{7}\}$.

CHECK

$$-7\left(-\frac{5}{7}\right) = 5$$

Since $0 \leqslant 5$, 0 is a solution; $0 \geqslant -\frac{5}{7}$ and is an element in the solution set.

16 **Solve $4x + 2 \geqslant 3x + 9$ and check your answer.**

$\{x \mid x \text{ is a real number and } x \geqslant 7\}$

CHECK

$$4(7) + 2 = 28 + 2 = 30 \qquad 3(7) + 9 = 21 + 9 = 30$$

Since $2 \geqslant 9$ is false, 0 is *not* a solution; 0 is not a member of the solution set.

17 Can you find a value of x which satisfies the inequation $x + 1 < x + 1$? What is the solution set of the inequation? Does the inequation have a boundary value?

---

no; $\emptyset$; no

---

18 Can you find a value of x which does *not* satisfy the inequation $x + 1 < x + 2$? What is the solution set of the inequation? Does the inequation have a boundary value?

---

no; $\{x \mid x \text{ is a real number}\}$; no

---

19 The inequation $5x - 3 > 2x - 7$ has a boundary value of $-\frac{4}{3}$ (Frame 13) and is a(n) (true/false/open) ______ sentence.

$x + 1 < x + 1$ is a(n) (true/false/open) ______ sentence.

$x + 1 < x + 2$ is a(n) (true/false/open) ______ sentence.

---

open; false; true

---

20 In order for an inequation to have a boundary value, it must be a(n) (true/false/open) ______ sentence.

---

open

---

21 A closed sentence can be shown to be equivalent to an equality or inequality. For example,

$$2x + 5 < 2x + 7 \longleftrightarrow 2x + (-2x) < 7 + (-5) \longleftrightarrow 0 < 2$$

which is a (true/false) ______ inequality; therefore the solution set is ______ ______.

---

true; $\{x \mid x \text{ is a real number}\}$

---

22 Similarly $3x - 7 > 3x - 6 \longleftrightarrow 3x + (-3x) > -6 + 7 \longleftrightarrow 0 > 1$ which is a ______ (true/false) inequality, and so the solution set is ______.

---

false; $\emptyset$

---

23 A single substitution is usually sufficient check for a closed sentence, as the solution set can often be determined by inspection. To check the answer to Frame 21, we substitute 0 for x in $2x + 5 < 2x + 7$.
This gives $0 + 5 < 0 + 7$, which is (true/false) ______, and so 0 (is/is not) ______ a solution, which (does/does not) ________ confirm our answer.

---

true; is; does

---

24 Check the answer to Frame 22 by a similar method.

Substituting 0 for x in $3x - 7 > 3x - 6$ gives $0 - 7 > 0 - 6$, which is false. Therefore 0 is not a solution, which confirms our answer.

---

25 Give the solution sets of the following inequations and check your answers. (Remember to check the boundary value when there is one.)

a. $4x + 3 > 5x + 4$
b. $3x + 7 \leqslant 3x + 2$
c. $2x - 4 < 2x - 2$

---

a. $\{x \mid x \text{ is a real number and } x < -1\}$

CHECK

$4(-1) + 3 = -4 + 3 = -1$ $\qquad$ $5(-1) + 4 = -5 + 4 = -1$

Since $0 + 3 > 0 + 4$ is false, 0 is not a solution; 0 is not a member of the solution set.

b. $\emptyset$

CHECK: Since $0 + 7 \leqslant 0 + 2$ is false, 0 is not a solution.

c. $\{x \mid x \text{ is a real number}\}$

CHECK: Since $0 - 4 < 0 - 2$ is true, 0 is a solution.

---

26 Give the solution sets of the following inequations and check your answers:

a. $2x + 4 \geqslant 2x + 2$
b. $2x + 3 \leqslant 4x + 7$
c. $2x - 9 > 2x + 1$

---

a. $\{x \mid x \text{ is a real number}\}$

CHECK: Since $0 + 4 \geqslant 0 + 2$ is true, 0 is a solution.

b. $\{x \mid x \text{ is a real number and } x \geqslant -2\}$

CHECK

$2(-2) + 3 = -4 + 3 = -1$
$4(-2) + 7 = -8 + 7 = -1$

Since $0 + 3 \leqslant 0 + 7$ is true, 0 is a solution; 0 is a member of the solution set.

c. $\emptyset$

CHECK: Since $0 - 9 > 0 + 1$ is false, 0 is not a solution.

You have finished the program for Sec. 3-3. Now take the posttest to see how much you have learned.

POSTTEST 3-3

Give the solution sets of the following inequations and check your answers:

1. $2x + 5 \geqslant 5x - 3$
2. $3x - 2 < 3x - 4$
3. $6x + 5 > 3x + 8$
4. $5x + 4 \geqslant 5x - 3$
5. $4x - 7 \leqslant 2x - 8$

When you have finished the posttest, correct it yourself using the answers on page 167. Then check all the questions you got wrong on the list below:

1. ______
2. ______
3. ______
4. ______
5. ______

If you compare your performance with that on Pretest 3-3, you will see how much you have improved. If you made more than one mistake on this posttest, review the material and take the pretest again. Don't move on to the next section until you can get all the questions right.

Now proceed to Final Test 3.

FINAL TEST 3

Give the solution sets of the following inequations

1. $3x - 5 \leqslant 5x + 9$
2. $-5x > 3$

Graph the solution sets of the following inequations on the number line:

3. $5y - 4 \geqslant 2y - 7$
4. $4y + 7 < 7y + 3$
5. $3x - 4 \neq 5x + 2$

Give the solution sets of the following inequations and check your answers:

6. $2y - 5 \geqslant 5y + 6$
7. $6y - 3 < 6y + 2$
8. $4y - 5 > 4y - 1$

The answers to this test are on page 168. Check all the questions you got wrong on the list below and compare your results with those on Diagnostic Test 3 to see how much you have improved.

1. _____	4. _____	7. _____
2. _____	5. _____	8. _____
3. _____	6. _____	

If you made more than one mistake, review the material and then take the diagnostic test again. Make sure that you can get all the questions right before you finish the chapter. Now proceed to Chap. 4.

ANSWERS TO TESTS FOR CHAPTER 3

DIAGNOSTIC TEST 3

1. $\{x \mid x \text{ is a real number and } x > -3\}$

2. $\{x \mid x \text{ is a real number and } x \leqslant -\frac{7}{3}\}$

3. 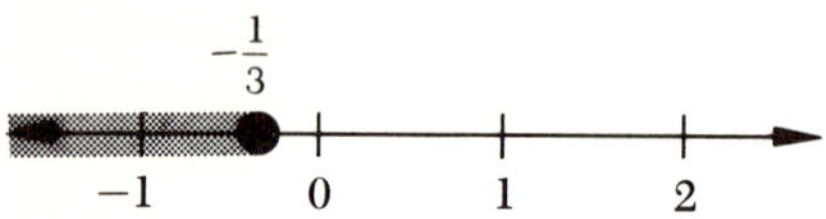

4. 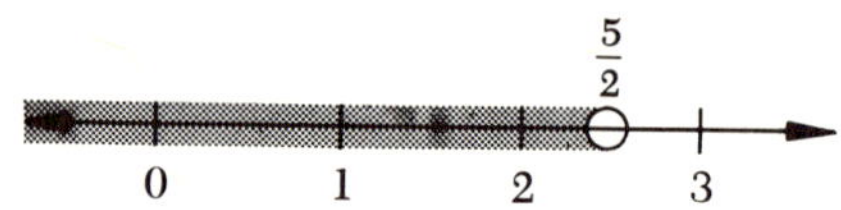

5. 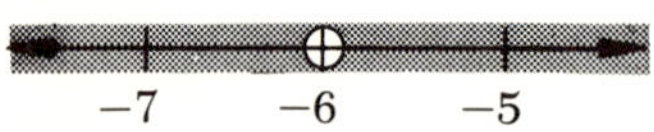

6. $\{x \mid x \text{ is a real number and } x \geqslant -\frac{11}{4}\}$

   CHECK

   $$3\left(-\frac{11}{4}\right) - 7 = \frac{-33 - 28}{4} = -\frac{61}{4} \qquad 7\left(-\frac{11}{4}\right) + 4 = \frac{-77 + 16}{4} = -\frac{61}{4}$$

   Substituting 0 for x, $0 - 7 \leqslant 4$ is true, and so 0 is a solution. [Do not count this question correct unless you substituted $-\frac{11}{4}$ for x and one other value (not necessarily 0).]

7. $\{x \mid x \text{ is a real number}\}$

   CHECK: Substituting 0 for x, $0 + 3 > -7$ is true, and so 0 is a solution. [Do not count this question correct unless you checked your answer by at least one substitution (not necessarily 0).]

8. $\emptyset$

   CHECK: Substituting 0 for x, $0 - 2 < 0 - 6$ is false, and so 0 is not a solution. [Do not count this question correct unless you checked your answer by at least one substitution (not necessarily 0).]

PRETEST 3-1

1. $\{x \mid x \text{ is a real number and } x \geqslant 2\}$

2. $\{x \mid x \text{ is a real number and } x < -\frac{3}{7}\}$

3. $\{x \mid x \text{ is a real number and } x \leqslant -\frac{5}{6}\}$

4. $\{x \mid x \text{ is a real number and } x \geqslant \frac{2}{5}\}$

5. $\{x \mid x \text{ is a real number and } x < -\frac{13}{5}\}$

POSTTEST 3-1

1. $\{x \mid x \text{ is a real number and } x \leqslant 1\}$
2. $\{x \mid x \text{ is a real number and } x > -\frac{8}{3}\}$
3. $\{x \mid x \text{ is a real number and } x \geqslant -\frac{4}{7}\}$
4. $\{x \mid x \text{ is a real number and } x \leqslant -\frac{3}{5}\}$
5. $\{x \mid x \text{ is a real number and } x > \frac{11}{5}\}$

PRETEST 3-2

1. 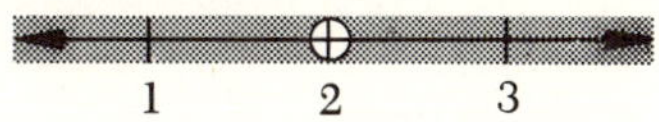

2. 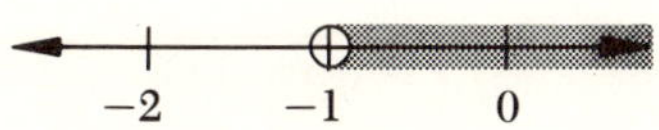

3. 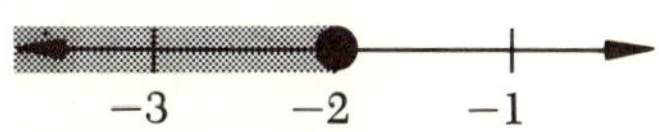

4. 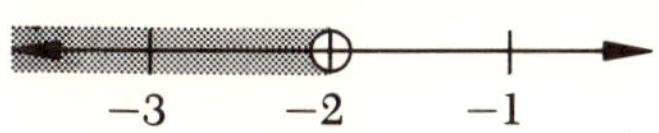

5. 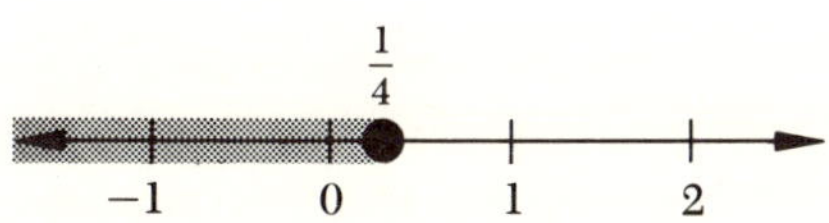

6. 

POSTTEST 3-2

1. 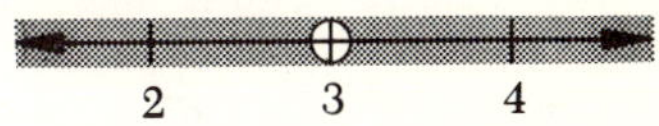

2. 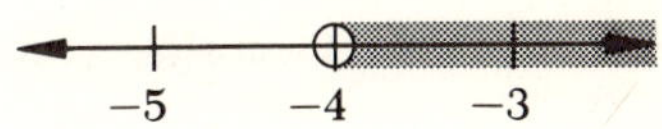

3. 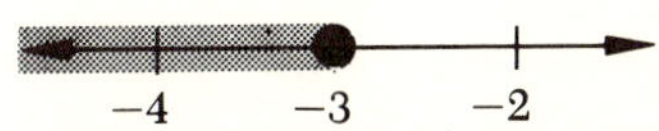

4. 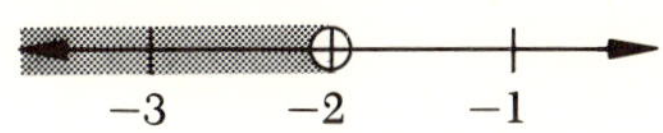

5. 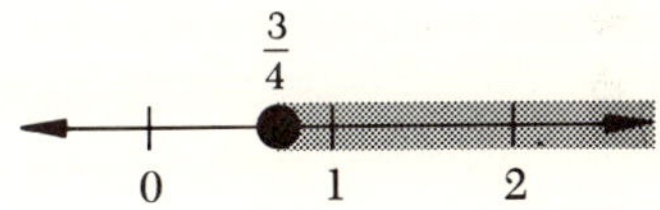

6. 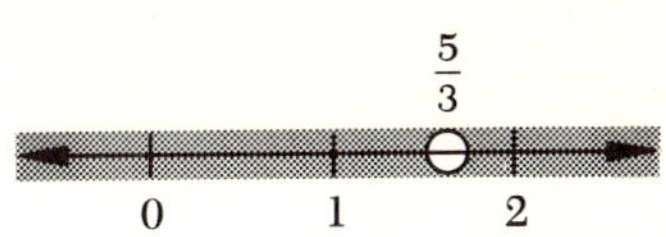

PRETEST 3-3

1. $\{x \mid x \text{ is a real number and } x \geqslant -\frac{7}{4}\}$

CHECK

$$3\left(-\frac{7}{4}\right) - 4 = \frac{-21 - 16}{4} = -\frac{37}{4} \qquad 7\left(-\frac{7}{4}\right) + 3 = \frac{-49 + 12}{4} = -\frac{37}{4}$$

Substituting 0 for x gives $0 - 4 \leqslant 0 + 3$, which is true, and so 0 is a solution. [Do not count this question correct unless you substituted $-\dfrac{7}{4}$ for x and one other value (not necessarily 0).]

2. $\{x \mid x \text{ is a real number}\}$

CHECK: Substituting 0 for x gives $0 + 7 > 0 + 5$, which is true, and so 0 is a solution. [Do not count this question correct unless you checked your answer by at least one substitution (not necessarily 0).]

3. $\{x \mid x \text{ is a real number and } x > -\frac{3}{4}\}$

CHECK

$$5\left(-\frac{3}{4}\right) + 2 = \frac{-15 + 8}{4} = -\frac{7}{4}$$

$$9\left(-\frac{3}{4}\right) + 5 = \frac{-27 + 20}{4} = -\frac{7}{4}$$

Substituting 0 for x gives $0 + 2 < 0 + 5$, which is true, and so 0 is a solution. [Do not count this question as correct unless you substituted $-\frac{3}{4}$ for x and one other value (not necessarily 0).]

4. $\emptyset$

CHECK: Substituting 0 for x gives $0 - 4 \geqslant 0 + 7$, which is false, and so 0 is not a solution. [Do not count this question correct unless you checked your answer by at least one substitution (not necessarily 0).]

5. $\{x \mid x \text{ is a real number and } x \geqslant \frac{2}{3}\}$

CHECK

$$6\left(\frac{2}{3}\right) - 5 = 4 - 5 = -1$$

$$3\left(\frac{2}{3}\right) - 3 = 2 - 3 = -1$$

Substituting 0 for x gives $0 - 5 \geqslant 0 - 3$, which is false, and so 0 is not a solution. [Do not count this question correct unless you substituted $\frac{2}{3}$ for x and one other value (not necessarily 0).]

POSTTEST 3-3

1. $\{x \mid x \text{ is a real number and } x \leqslant \frac{8}{3}\}$

   CHECK

   $$2\left(\frac{8}{3}\right) + 5 = \frac{16 + 15}{3} = \frac{31}{3} \qquad 5\left(\frac{8}{3}\right) - 3 = \frac{40 - 9}{3} = \frac{31}{3}$$

   Substituting 0 for x gives $0 + 5 \geqslant 0 - 3$, which is true, and so 0 is a solution. [Do not count this question as correct unless you substituted $\frac{8}{3}$ for x and one other value (not necessarily 0).]

2. $\emptyset$

   CHECK: Substituting 0 for x gives $0 - 2 < 0 - 4$, which is false, and so 0 is not a solution. [Do not count this question as correct unless you checked your answer by at least one substitution (not necessarily 0).]

3. $\{x \mid x \text{ is a real number and } x > 1\}$

   CHECK

   $6(1) + 5 = 11$
   $3(1) + 8 = 11$

   Substituting 0 for x gives $0 + 5 > 0 + 8$, which is false, and so 0 is not a solution. [Do not count this question as correct unless you substituted 1 for x and one other value (not necessarily 0).]

4. $\{x \mid x \text{ is a real number}\}$

   CHECK: Substituting 0 for x gives $0 + 4 \geqslant 0 - 3$, which is true, and so 0 is a solution. [Do not count this question as correct unless you checked your answer by at least one substitution (not necessarily 0).]

5. $\{x \mid x \text{ is a real number and } x \leqslant -\frac{1}{2}\}$

   CHECK

   $$4\left(-\frac{1}{2}\right) - 7 = -2 - 7 = -9$$

   $$2\left(-\frac{1}{2}\right) - 8 = -1 - 8 = -9$$

   Substituting 0 for x gives $0 - 7 \leqslant 0 - 8$, which is false, and so 0 is not a solution. [Do not count this question as correct unless you substituted $-\frac{1}{2}$ for x and one other value (not necessarily 0).]

FINAL TEST 3

1. $\{x \mid x \text{ is a real number and } x \geqslant -7\}$

2. $\{x \mid x \text{ is a real number and } x < -\frac{3}{5}\}$

3. 

4. 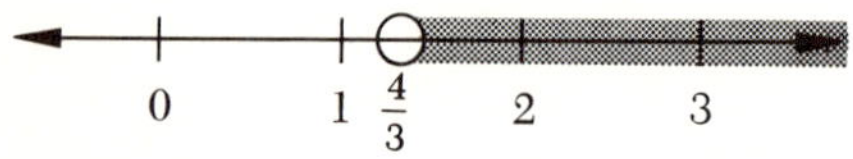

5. 

6. $\{y \mid y \text{ is a real number and } y \leqslant -\frac{11}{3}\}$

CHECK

$$2\left(-\frac{11}{3}\right) - 5 = \frac{-22 - 15}{3} = -\frac{37}{3}$$

$$5\left(-\frac{11}{3}\right) + 6 = \frac{-55 + 18}{3} = -\frac{37}{3}$$

Substituting 0 for y, $-5 \geqslant 6$ is false, so 0 is not a solution. [Do not count this question correct unless you substituted $-\frac{11}{3}$ for y and one other value (not necessarily 0).]

7. $\{y \mid y \text{ is a real number}\}$

CHECK: Substituting 0 for y, $-3 < 2$ is true, so 0 is a solution. [Do not count this question correct unless you checked your answer by at least one substitution (not necessarily 0).]

8. $\emptyset$

CHECK: Substituting 0 for y, $-5 > -1$ is false, so 0 is not a solution. [Do not count this question correct unless you checked your answer by at least one substitution (not necessarily 0).]

Chapter 4 Equations Involving Absolute Value

The absolute value of an expression is introduced in Sec. 4-1 as a notation for expressing the distance between two points on a number line. You are shown how to use this definition to solve equations of the type $|y - k| = b$. Section 4-2 extends its use to the solution of equations of the type $|ay - k| = b$. Section 4-3 explains how to graph these equations and the corresponding inequations. Throughout the chapter the domain of the variable is assumed to be the set of real numbers.

This chapter requires knowledge of Chaps. 2 and 3.

DIAGNOSTIC TEST 4

Give the solution sets of the following equations:

1. $|y + 2| = 4$
2. $|w| = 8$
3. $|z - 6| = -1$
4. $|3y + 7| = 4$
5. $|5y - 1| = 3$

Graph the solution sets of the following open sentences:

6. $|y - 4| = 1$
7. $|6x + 2| \leqslant 5$
8. $|4x - 3| > 2$

(You are advised to check your answers.)

When you have finished the diagnostic test, correct it yourself using the answers on page 192. Then check all the questions you got wrong on the list below:

1. ______	4. ______	6. ______
2. ______	5. ______	7. ______
3. ______		8. ______

If you made no mistakes at all, you need not do any work in this chapter and should go straight on to Chap. 5, First-degree Equations in Two Variables, which begins on page 195. But if you made any mistakes (even if it was only *one* mistake), follow these instructions.

	section	page
If you made any mistakes in Questions 1 to 3, circle Sec. 4-1.	4-1	171
If you made any mistakes in Question 4 or 5, circle Sec. 4-2.	4-2	178
If you made any mistakes in Questions 6 to 8, circle Sec. 4-3.	4-3	186

You should now work through the sections whose numbers you have circled, taking them in numerical order. The sections begin on the pages indicated.

Section 4-1 Solving Equations of the form $|y - k| = b$

Do this section if you made a mistake in Questions 1 to 3 of Diagnostic Test 4. Begin by taking the pretest below, which will show you how much of Sec. 4-1 you already know and what you will be able to do by the time you have finished it.

PRETEST 4-1

Give the solution sets of the following equations:

1. $|y - 2| = 7$
2. $|y + 5| = 3$
3. $|z| = 6$
4. $|y + 3| = 0$
5. $|k - 2| = -3$

(You are advised to check your answers.)

When you have finished the pretest, correct it yourself using the answers on page 193. Then check all the questions you got wrong on the list below:

1. ______ 4. ______

2. ______ 5. ______

3. ______

If you made any mistakes (even if it was only *one* mistake), read the introduction and go through the program.

If you made no mistakes at all, skip Sec. 4-1, because you already know its material, and go on to the next section whose number you circled on page 170. If no further numbers were circled, take Final Test 4 (page 191).

Introduction

You will learn how to use the absolute value of an expression to represent the distance between two points A and B on a number line when it is not known whether $A < B$ or $A > B$. Like distance, absolute value is defined so that it is never negative. You are shown that the general equation $|y - k| = b$ signifies that y is at a distance b from the point k. The equation has two solutions when $b > 0$, one when $b = 0$, and none when $b < 0$. The cases $k > 0$, $k = 0$, and $k < 0$ are all considered.

FRAMES FOR SECTION 4-1

1 *a.* What is the distance between points 1 and 3 on the number line?
b. What is the distance between points 5 and 3 on the number line?
c. Will the distance between two points always be a positive number?

---

a. 2 *b.* 2 *c.* yes

---

2 Suppose we have two points A and B on a number line. If A is to the right of B, that is, $A > B$,

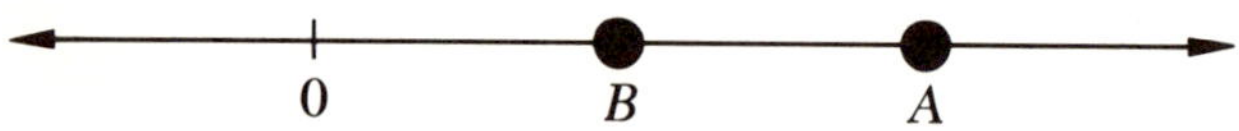

the distance between A and B will be ______ since a distance must always be positive.

$A - B$

---

3 But if A is to the left of B, that is, $A < B$,

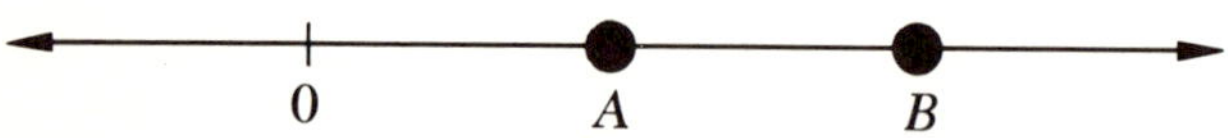

the distance between A and B will be ______. (Remember that it must be positive.)

$B - A$

---

4 If we do not know the relative positions of A and B, we cannot describe the distance between them as $A - B$ or as $B - A$ because either description could be false. Instead we describe it as $|A - B|$, the *absolute value* of $A - B$, which equals either $A - B$ or $-(A - B) = B - A$, whichever of the two is positive. Therefore:

a. If $A > B$, $|A - B| =$ ______.

b. And if $A < B$, $|A - B| =$ ______.

---

a. $A - B$ *b.* $-(A - B)$ or $B - A$

---

5 How would you describe the distance between points C and D on the number line (*a*) if C is greater than D? (*b*) if C is less than D? (*c*) if you don't know whether C is greater or less than D?

a. $C - D$ *b.* $D - C$ *c.* $|C - D|$ or $|D - C|$

6 Since $|C - D|$ is either $C - D$ or $-(C - D)$, whichever is positive, it follows that $|n|$ is either ______ or ______, whichever is the positive.

n $-n$

7 Hence

a. $|3| =$ ______ and $|-3| =$ ______.

b. $|9| =$ ______ and $|-9| =$ ______.

a. 3; 3 *b.* 9; 9

8 *a.* $|x - 3|$ is the distance on the number line between point x and point ______.

b. If $x = 2$, this distance equals $|2 - 3| = |-1| =$ ______.

c. If $x = 5$, this distance equals $|\_ - 3| = |\_| =$ ______.

d. If $x = 0$, this distance equals $|\_ - 3| = |\_| =$ ______.

e. If $x = -2$, this distance equals $|\_ - 3| = |\_| =$ ______.

a. 3 *b.* 1 *c.* $|5 - 3| = |2| = 2$ *d.* $|0 - 3| = |-3| = 3$
e. $|-2 - 3| = |-5| = 5$

9 If $|x - 3| = 4$, we know that the distance between point x and point ______ is equal to ______

3 4

10 Look at point 3 on the number line below.

a. What points on the line are at a distance of 4 from the point 3?

b. What values of x satisfy the equation $|x - 3| = 4$?

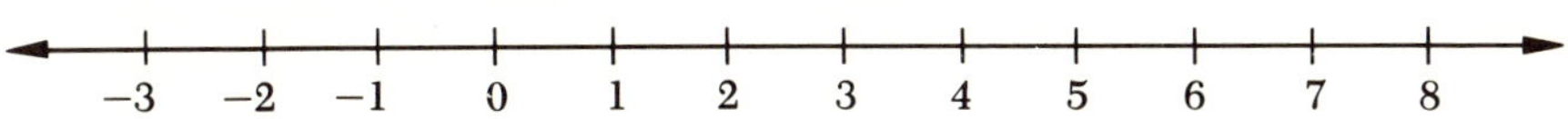

a. −1 and 7 *b.* −1 and 7

11 Thus the solution set of the equation $|x - 3| = 4$ is {______, ______}. Check both solutions in the original equation.

$\{-1, 7\}$

CHECK

$|-1 - 3| = |-4| = 4$
$|7 - 3| = |4| = 4$

---

12 If $|x - 5| = 2$, we know that the distance between point x and point ______ is equal to ______. How many values of x satisfy this condition?

---

5 2 two values (x can be to the left or right of 5)

---

13 Give the solution set of the equation $|x - 5| = 2$ and check your answer.

$\{3, 7\}$

CHECK

$|3 - 5| = |-2| = 2$
$|7 - 5| = |2| = 2$

---

14 Give the solution set of $|x - 1| = 7$ and check your answer.

$\{-6, 8\}$

CHECK

$|-6 - 1| = |-7| = 7$
$|8 - 1| = |7| = 7$

---

15 Give the solution set of $|x - 7| = 1$ and check your answer.

$\{6, 8\}$

CHECK

$|-6 - 1| = |-7| = 7$
$|8 - 1| = |7| = 7$

---

16 If $|x + 5| = 3$, we know that $|x - (-5)| = 3$ and therefore that the distance between point x and point ______ is equal to ______. Give the solution set of $|x + 5| = 3$ and check your answer.

-5 3 $\{-8, -2\}$

CHECK

$|-8 + 5| = |-3| = 3$
$|-2 + 5| = |3| = 3$

---

17 Give the solution set of $|x + 2| = 5$ and check your answer.

$\{-7, 3\}$

CHECK

$|-7 + 2| = |-5| = 5$
$|3 + 2| = |5| = 5$

---

18 Give the solution set of $|x + 3| = 7$ and check your answer.

$\{-10, 4\}$

CHECK

$|-10 + 3| = |-7| = 7$
$|4 + 3| = |7| = 7$

---

19 *a.* $|6| =$ ______; $|-6| =$ ______.
b. Can the absolute value of any number be negative?
c. Does $|y| = -3$ have any solutions?

---

a. 6; 6 *b.* no *c.* no

---

20 *a.* Give the solution set of $|y| = -7$.
b. Give the solution set of $|y| = 7$.

---

a. $\emptyset$ *b.* $\{-7, 7\}$

---

21 *a.* Give the solution set of $|z| = 3$.
b. Give the solution set of $|w| = -2$.

---

a. $\{-3, 3\}$ *b.* $\emptyset$

---

22 $|z| = 3$ can be said to represent a situation in which z is located at a distance of 3 from the point ______ since it can also be written as $|z - \_\_\_\_\_\_| = 3$.

---

0 0

---

23 What equation would you use to describe a situation in which z was located at a distance of 0 from the point 3? What would be its solution set?

$|z - 3| = 0$ $\{3\}$

---

24 What equation would you use to describe a situation in which z was located at a distance of 0 from the point -3? What would be its solution set?

$|z + 3| = 0$ $\{-3\}$

---

25 The equation $|z - 5| = k$ has:

a. ______ solution(s) if $k < 0$
b. ______ solution(s) if $k = 0$
c. ______ solution(s) if $k > 0$

---

a. no *b.* one *c.* two

---

26 Give the solution set of:

a. $|x - 1| = 5$
b. $|x - 1| = -5$
c. $|x - 1| = 0$

---

a. $\{-4, 6\}$ *b.* $\emptyset$ *c.* $\{1\}$

---

27 Give the solution set of:

a. $|x + 1| = 5$
b. $|x + 1| = -5$
c. $|x + 1| = 0$

---

a. $\{-6, 4\}$ *b.* $\emptyset$ *c.* $\{-1\}$

---

28 Give the solution set of:

a. $|x| = 5$
b. $|x| = -5$
c. $|x| = 0$

---

a. $\{-5, 5\}$ *b.* $\emptyset$ *c.* $\{0\}$

You have finished the program for Sec. 4-1. Now take the posttest to see how much you have learned.

POSTTEST 4-1

Give the solution sets of the following equations:

1. $|y - 3| = 4$
2. $|y + 2| = 5$
3. $|k| = 7$
4. $|y + 6| = 0$
5. $|k - 3| = -4$

(You are advised to check your answers.)

When you have finished the posttest, correct it yourself using the answers on page 193. Then check all the questions you got wrong on the list below:

1. ______ 4. ______
2. ______ 5. ______
3. ______

If you compare your performance with that on Pretest 4-1, you will see how much you have improved. If you made more than one mistake on this posttest, review the material and take the pretest again. Don't move on to the next section until you can get all the questions right.

Now proceed to the next section whose number you circled on page 170. If no other sections were circled, take Final Test 4 (page 191).

Section 4-2
Solving Equations of the Form $|ay - k| = b$

Do this section if you made a mistake in Question 4 or 5 of Diagnostic Test 4. Begin by taking the pretest below, which will show you how much of Sec. 4-2 you already know and what you will be able to do by the time you have finished it.

PRETEST 4-2

1. The solution set of $|3k - 5| = 6$ is the union of the solution sets of the equations ______ and ______.

Give the solution sets of the following equations:

2. $|2w + 3| = 7$
3. $|3w - 5| = 1$
4. $|4w + 6| = 3$
5. $|5w - 2| = 4$

(You are advised to check your answers.)

When you have finished the pretest, correct it yourself using the answers on page 193. Then check all the questions you got wrong on the list below:

1. ______
2. ______
3. ______
4. ______
5. ______

If you made any mistakes (even if it was only *one* mistake), read the introduction and go through the program.

If you made no mistakes at all, skip Sec. 4-2, because you already know its material, and go on to the next section whose number you circled on page 170. If no further numbers were circled, take Final Test 4 (page 191).

Introduction

You are shown that the solution set of $|y - k| = b$ is the union of the solution sets of $y - k = b$ and $y - k = -b$, and you learn to use this property to solve equations of the form $|ay - k| = b$. You will see that these equations can also be solved by the distance method of Sec. 4-1 if they are multiplied by $1/a$ to give $|y - k/a| = b/a$. The two methods can be combined by solving $ay - k = \pm b$.

FRAMES FOR SECTION 4-2

1 If $|z| = 4$, then $z =$ ______ or ______, and if $|z + 1| = 3$, then $z + 1 =$ ______ or ______.

-4 4 -3 3

---

2 If $z + 1 = -3$, then $z =$ ______; and if $z + 1 = 3$, then $z =$ ______.

-4 2

---

3 Give the solution set of:

a. $z + 1 = -3$
b. $z + 1 = 3$
c. $|z + 1| = 3$

---

a. $\{-4\}$ *b.* $\{2\}$ *c.* $\{-4, 2\}$

---

4 The solution set of $|z + 1| = 3$ is the union of the solution sets of ______ and ______.

$z + 1 = -3$ $z + 1 = 3$

---

5 Give the solution set of:

a. $z - 5 = 4$
b. $z - 5 = -4$

---

a. $\{9\}$ *b.* $\{1\}$

---

6 Give an equation whose solution set is $\{1, 9\}$

$|z - 5| = 4$

---

7 Give the solution set of:

a. $3z - 4 = 5$
b. $3z - 4 = -5$
c. $|3z - 4| = 5$

---

a. $\{3\}$ *b.* $\{-\frac{1}{3}\}$ *c.* $\{-\frac{1}{3}, 3\}$

---

8 *a.* $5z - 3 = 7 \longleftrightarrow 5z =$ ______ $\longleftrightarrow z =$ ______

b. $5z - 3 = -7 \longleftrightarrow 5z =$ ______ $\longleftrightarrow z =$ ______

c. The solution set of $|5z - 3| = 7$ is ______. Check your answer.

---

a. 10; 2 *b.* $-4; -\frac{4}{5}$ *c.* $\{-\frac{4}{5}; 2\}$

CHECK

$$|5(-\tfrac{4}{5}) - 3| = |-4 - 3| = |-7| = 7$$
$$|5(2) - 3| = |10 - 3| = |7| = 7$$

---

9 Give the solution set of $|4y - 1| = 2$ and check your answer.

$4y - 1 = 2$ and $4y - 1 = -2$

$4y = 3$ $4y = -1$

$y = \frac{3}{4}$ $y = -\frac{1}{4}$

$$\left\{-\frac{1}{4}, \frac{3}{4}\right\}$$

CHECK

$$\left|4\left(-\frac{1}{4}\right) - 1\right| = |-1 - 1| = 2$$

$$\left|4\left(\frac{3}{4}\right) - 1\right| = |3 - 1| = 2$$

---

10 Give the solution set of $|3y + 5| = 6$ and check your answer.

$3y + 5 = 6$ and $3y + 5 = -6$

$3y = 1$ $3y = -11$

$y = \frac{1}{3}$ $y = -\frac{11}{3}$

$$\left\{-\frac{11}{3}, \frac{1}{3}\right\}$$

CHECK

$$\left|3\left(-\frac{11}{3}\right) + 5\right| = |-11 + 5| = 6$$

$$\left|3\left(\frac{1}{3}\right) + 5\right| = |1 + 5| = 6$$

---

11 An alternative method of solving this equation uses the method of the previous section:

$$|3y + 5| = 6$$

$$\Big\updownarrow \text{ MT with } \tfrac{1}{3}$$

$$\left|y + \frac{5}{3}\right| = 2$$

This signifies that y is at a distance of ______ from the point $-\frac{5}{3}$ and therefore $y = -\frac{5}{3} +$ ______ or $-\frac{5}{3} -$ ______ = ______ or ______. Compare this result with your answer to Frame 10.

---

2 2 2 $\frac{1}{3}$ or $-\frac{11}{3}$

---

12 $|2y + 7| = 8$

$$\Big\updownarrow \text{ MT with } \tfrac{1}{2}$$

$$\left|y + \frac{7}{2}\right| = \text{\_\_\_\_\_\_}$$

which signifies that y is at a distance of ______ from the point ______ and therefore that $y =$ ______ + ______ or ______ − ______ = ______ or ______. Give the solution set of $|2y + 7| = 8$ and check your answer in the original equation.

---

4 4 $-\frac{7}{2}$ $-\frac{7}{2} + 4$ $-\frac{7}{2} - 4 = \frac{1}{2}$ or $-\frac{15}{2}$ $\left\{-\frac{15}{2}, \frac{1}{2}\right\}$

CHECK

$$\left|2\left(-\frac{15}{2}\right) + 7\right| = |-15 + 7| = 8$$

$$\left|2\left(\frac{1}{2}\right) + 7\right| = |1 + 7| = 8$$

---

13 Use the method of Frames 11 and 12 to find the solution set of $|4y - 6| = 3$. Check your answer in the original equation

$$|4y - 6| = 3$$

$$\updownarrow \text{MT with } \tfrac{1}{4}$$

$$\left|y - \frac{6}{4}\right| = \frac{3}{4}$$

$$y = \frac{6}{4} + \frac{3}{4} = \frac{9}{4} \quad \text{or} \quad \frac{6}{4} - \frac{3}{4} = \frac{3}{4}$$

$$\left\{\frac{3}{4}, \frac{9}{4}\right\}$$

CHECK

$$\left|4\left(\frac{3}{4}\right) - 6\right| = |3 - 6| = 3$$

$$\left|4\left(\frac{9}{4}\right) - 6\right| = |9 - 6| = 3$$

---

14 The two methods we have used are substantially the same. The first method used the fact that $|4y - 6| = 3$ implies that either $4y - 6 = 3$ or $4y - 6 = -3$, which can be abbreviated to $4y - 6 = \pm 3$ ($\pm$ means $+$ or $-$). If we now proceed as in the second method, we get

$$4y - 6 = \pm 3$$

$$\updownarrow \text{MT with } \tfrac{1}{4}$$

$$y - \frac{6}{4} = \pm\frac{3}{4}$$

$$\updownarrow \text{AT with } \tfrac{6}{4}$$

$$y = \underline{\quad\quad} \pm \underline{\quad\quad} \longleftrightarrow y = \underline{\quad\quad} \text{ or } \underline{\quad\quad}$$

Compare this result with your answer to Frame 13.

$$\frac{6}{4} \quad \frac{3}{4} \quad \frac{9}{4} \quad \frac{3}{4}$$

---

15 Using the same approach, we have

$$|5y - 3| = 2 \longleftrightarrow 5y - 3 = \pm 2$$

$\updownarrow$ MT with $\frac{1}{5}$

$$y - \frac{3}{5} = \_\_\_\_$$

$\updownarrow$ AT with $\frac{3}{5}$

$$y = \_\_\_\_$$

The solution set of $|5y - 3| = 2$ is ____. Check this answer.

$\pm\frac{2}{5}$ $\quad y = \frac{3}{5} \pm \frac{2}{5}$

$\left\{\frac{1}{5}, 1\right\}$

CHECK

$$\left|5\left(\frac{1}{5}\right) - 3\right| = |1 - 3| = 2$$

$$|5(1) - 3| = |5 - 3| = 2$$

---

16 Use the method of Frame 15 to find the solution set of $|2z - 3| = 5$ and check your answer.

$$2z - 3 = \pm 5$$

$$z - \frac{3}{2} = \pm\frac{5}{2}$$

$$z = \frac{3}{2} \pm \frac{5}{2}$$

$\{-1, 4\}$

CHECK

$$|2(-1) - 3| = |-2 - 3| = 5$$

$$|2(4) - 3| = |8 - 3| = 5$$

---

17 Give the solution set of $|3z + 4| = 1$ and check your answer.

$$3z + 4 = \pm 1$$

$$z + \frac{4}{3} = \pm\frac{1}{3}$$

$$z = -\frac{4}{3} \pm \frac{1}{3}$$

$$\left\{-\frac{5}{3}, -1\right\}$$

CHECK

$$\left|3\left(-\frac{5}{3}\right) + 4\right| = |-5 + 4| = 1$$

$$|3(-1) + 4| = |-3 + 4| = 1$$

18 Give the solution set of $|6k + 2| = 3$ and check your answer.

$$6k + 2 = \pm 3$$

$$k + \frac{2}{6} = \pm\frac{3}{6}$$

$$k = -\frac{2}{6} \pm \frac{3}{6}$$

$$\left\{-\frac{5}{6}, \frac{1}{6}\right\}$$

CHECK

$$\left|6\left(-\frac{5}{6}\right) + 2\right| = |-5 + 2| = 3$$

$$\left|6\left(\frac{1}{6}\right) + 2\right| = |1 + 2| = 3$$

You have finished the program for Sec. 4-2. Now take the posttest to see how much you have learned.

POSTTEST 4-2

1. The solution set of $|5k + 3| = 7$ is the union of the solution sets of the equations ______ and ______.

Give the solution sets of the following equations:

2. $|2w - 5| = 1$
3. $|3w + 2| = 6$
4. $|4w - 3| = 5$
5. $|5w + 7| = 4$

(You are advised to check your answers.)

When you have finished the posttest, correct it yourself using the answers on page 193. Then check all the questions you got wrong on the list below:

1. ______
2. ______
3. ______
4. ______
5. ______

If you compare your performance with that on Pretest 4-2, you will see how much you have improved. If you made more than one mistake on this post-test, review the material and take the pretest again. Don't move on to the next section until you can get all the questions right.

Now proceed to the next section whose number you circled on page 170. If no other sections were circled, take Final Test 4 (page 191).

Section 4-3
Solving and Graphing Equations and Inequations Involving Absolute Value

Do this section if you made a mistake in Questions 6 to 8 of Diagnostic Test 4. Begin by taking the pretest below, which will show you how much of Sec. 4-3 you already know and what you will be able to do by the time you have finished it.

PRETEST 4-3

Graph the solution sets of the following open sentences on the number line:

1. $|4x - 7| = 5$
2. $|x - 5| > 3$
3. $|x + 3| \leqslant 2$
4. $|2x - 7| < 5$
5. $|3x + 5| \geqslant 7$

When you have finished the pretest, correct it yourself using the answers on page 193. Then check all the questions you got wrong on the list below:

1. ______ 4. ______
2. ______ 5. ______
3. ______

If you made any mistakes (even if it was only *one* mistake), read the introduction and go through the program.

If you made no mistakes at all, skip Sec. 4-3, because you already know its material, and go on to take Final Test 4 (page 191).

Introduction

You will learn how to graph equations of the form $|ay - k| = b$ and hence also inequations of the form $|ay - k| < b$, $\leqslant b$, $> b$, and $\geqslant b$.

FRAMES FOR SECTION 4-3

1 Graph the solution sets of the following open sentences:

a. $x - 2 = 3$
b. $x - 2 < 3$
c. $x - 2 \geqslant 3$

---

a. 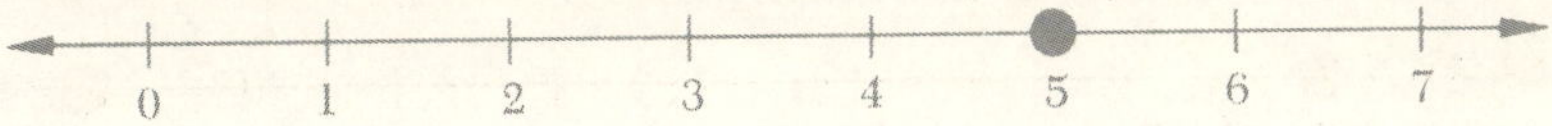

b. 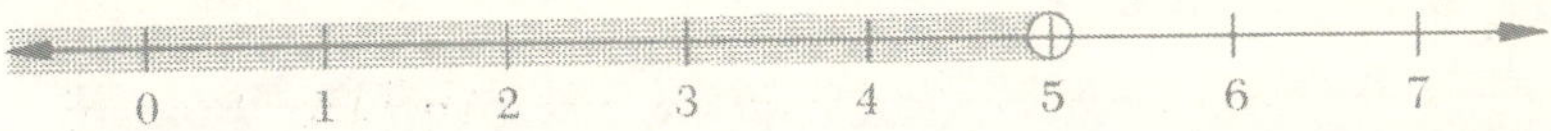

c. 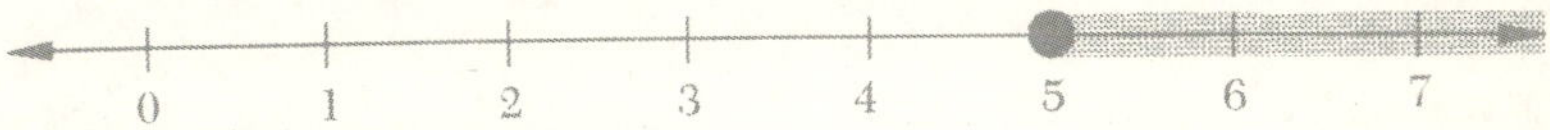

---

2 Graph the solution sets of the following open sentences:

a. $x + 1 = 2$
b. $x + 1 = -2$
c. $|x + 1| = 2$

---

a. 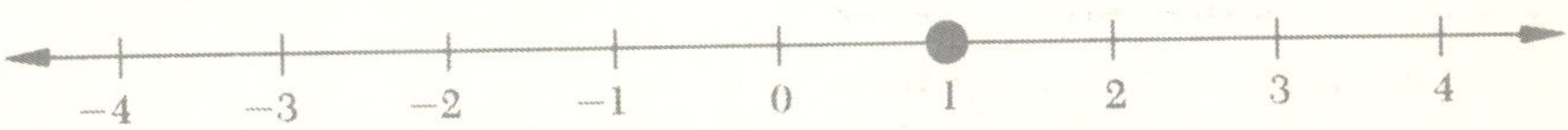

b. 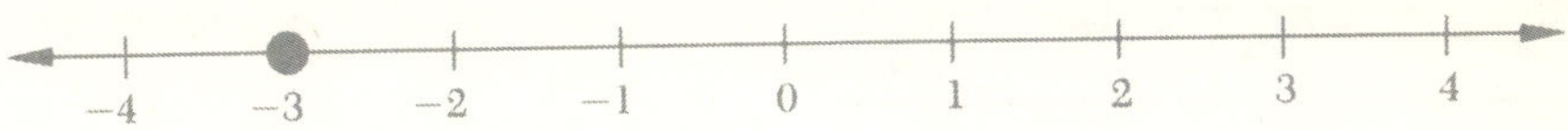

c. 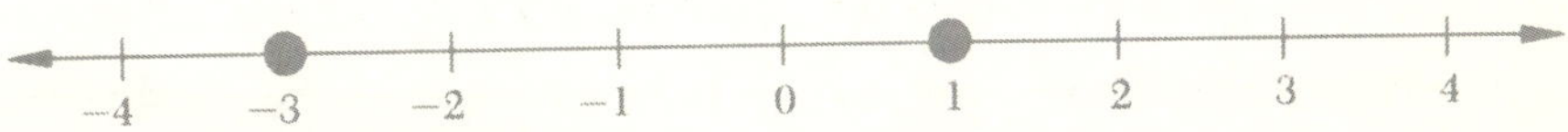

---

3 *a.* $|x + 1| = 2$ signifies that x is at a distance of ______ from the point ______.

b. Therefore $|x + 1| > 2$ signifies that x is at a distance greater than ______ from the point ______.

---

a. 2; −1 *b.* 2; −1

---

4 Which of the following is the graph of $|x + 1| > 2$?

a. 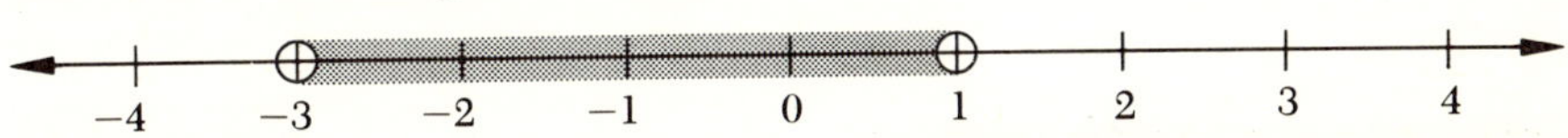

b. 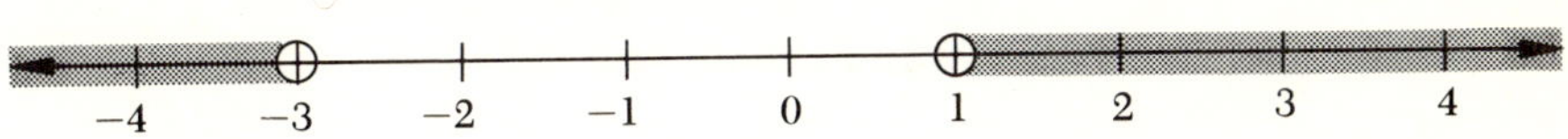

c. 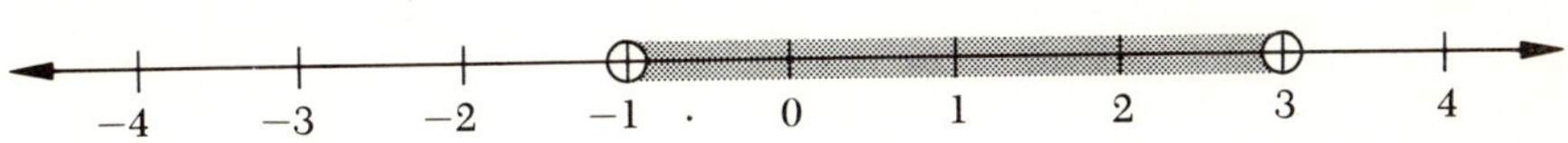

d. 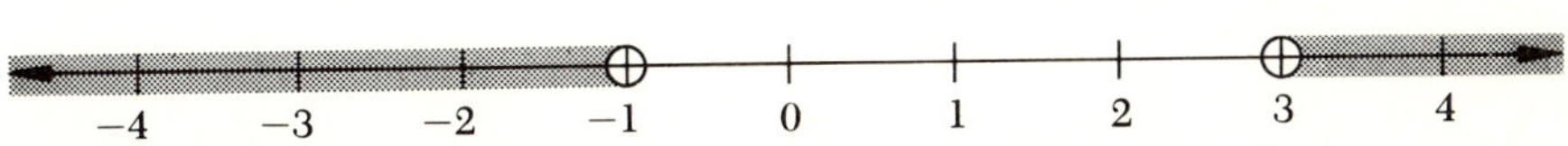

---

b

---

5 Which of the following is the graph $|x - 3| > 4$?

a. 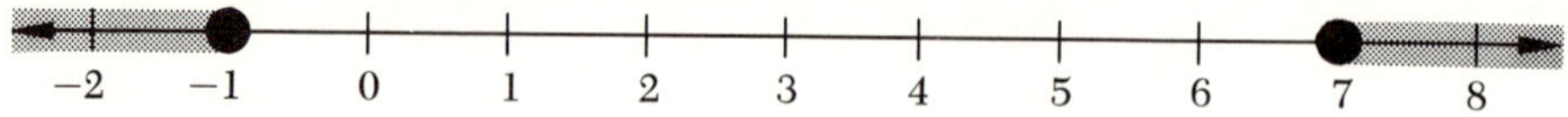

b. 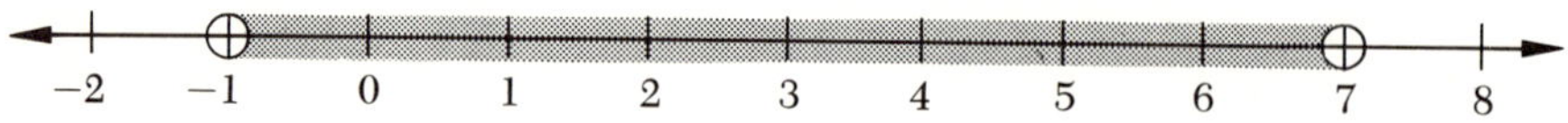

c. 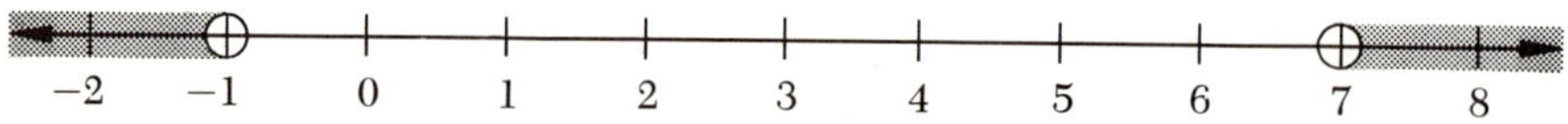

d. 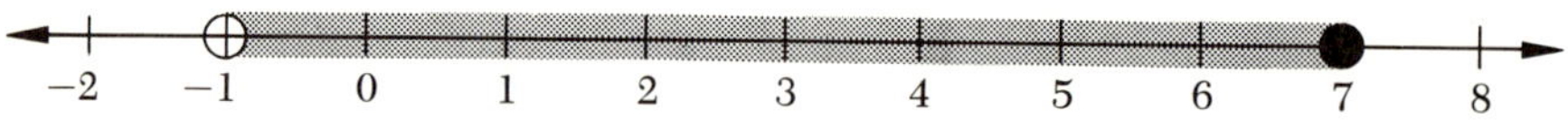

---

c

---

6 What is the solution set of $|x + 2| = 3$?

$\{-5, 1\}$

---

7 Which of the following is the graph $|x + 2| < 3$?

a. 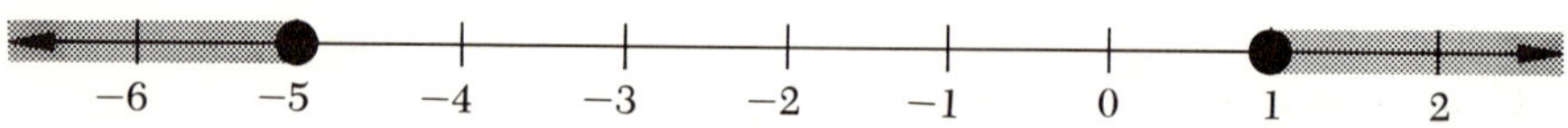

b. 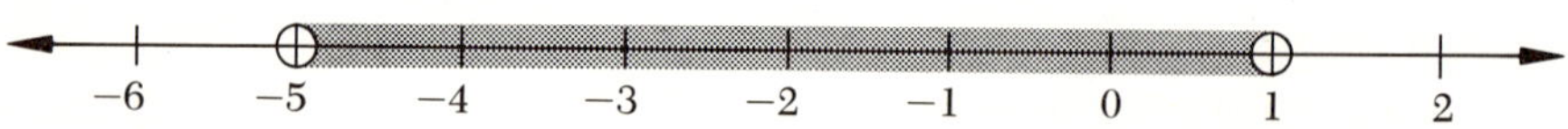

c. 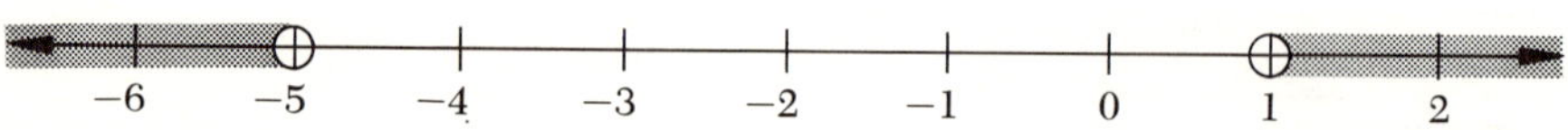

d. 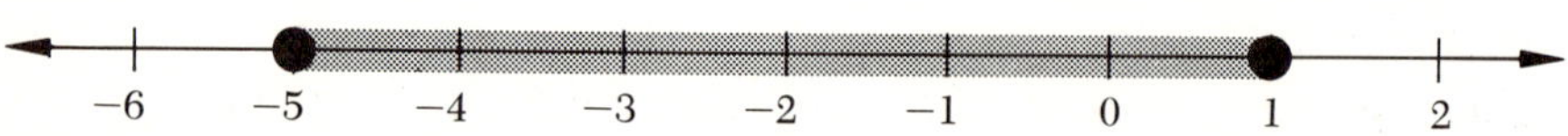

---

b

---

8 Graph the solution set of $|2x + 1| = 5$.

---

9 Graph the solution sets of:

a. $|2x + 1| > 5$
b. $|2x + 1| < 5$

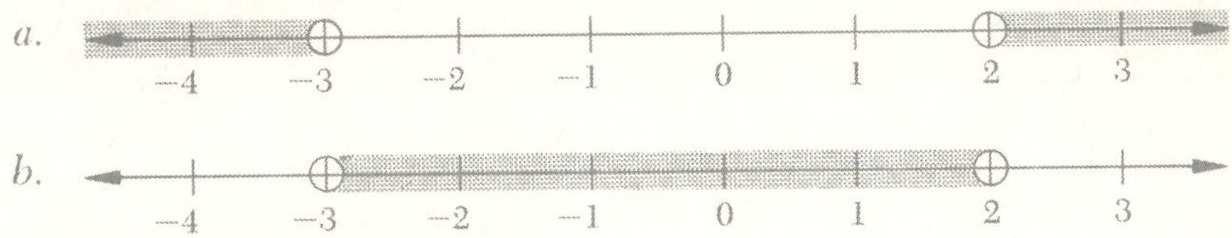

10 Graph the solution sets of

a. $|2x + 1| \geqslant 5$
b. $|2x + 1| \leqslant 5$

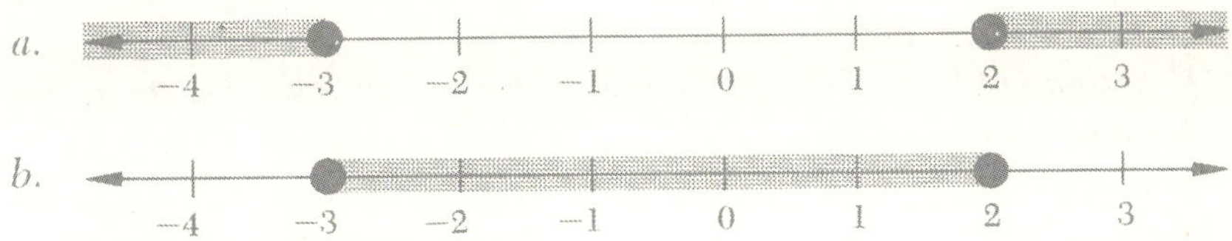

11 Graph the solution sets of:

a. $|2x - 3| = 1$
b. $|2x - 3| \geqslant 1$

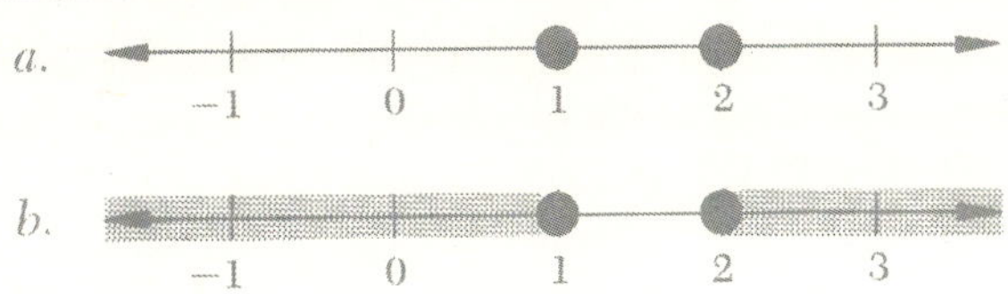

12 Graph the solution sets of:

a. $|3x + 4| = 7$
b. $|3x + 4| \leqslant 7$

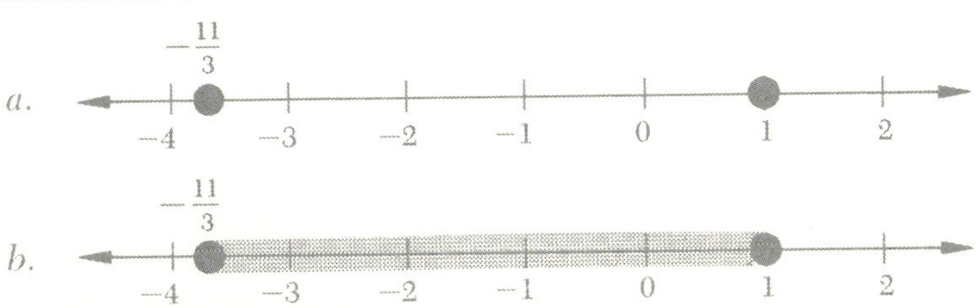

13 Graph the solution sets of:

a. $|x - 2| = 4$
b. $|x - 2| < 4$
c. $|x - 2| > 4$

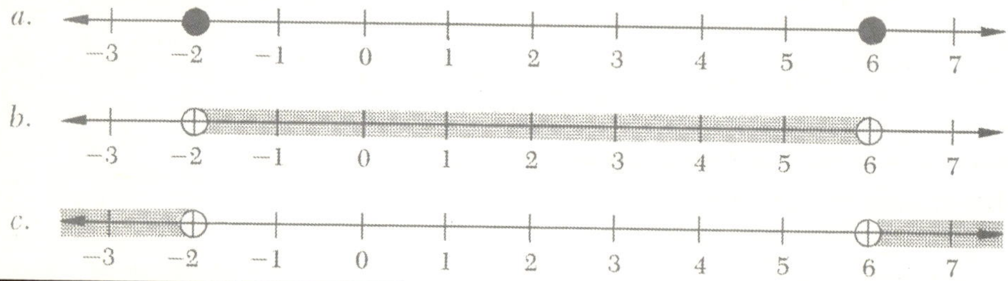

14 *a.* The graph of $|x - 2| = 4$ contains those points whose distance from the point 2 is equal to ______.
b. The graph of $|x - 2| < 4$ contains those points whose distance from the point 2 is ____________.
c. The graph of $|x - 2| > 4$ contains those points whose distance from the point 2 is ____________.

a. 4 *b.* less than 4 *c.* greater than 4

15 The graph of $|x - a| = b$ contains those points whose distance from the point a is equal to b. The graph of $|x - a| < b$ contains those points whose distance from the point a is ____________, i.e., all points (inside/outside) ______ the two solutions of $|x - a| = b$. The graph of $|x - a| > b$ contains those points whose distance from the point a is ____________, i.e., all points (inside/outside) ____________ the two solutions of $|x - a| = b$.

less than b; inside; greater than b; outside

16 Graph the solution sets of:

a. $|4x + 3| < 9$
b. $|3x + 4| \geq 10$

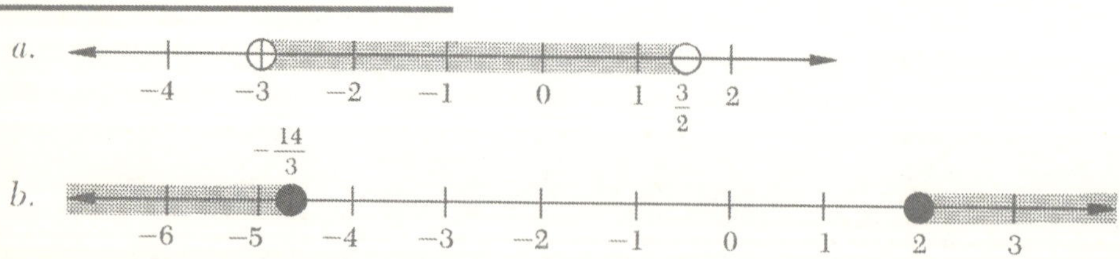

17 Graph the solution sets of:

a. $|2x - 7| > 4$
b. $|3x - 5| \leqslant 3$

---

a.

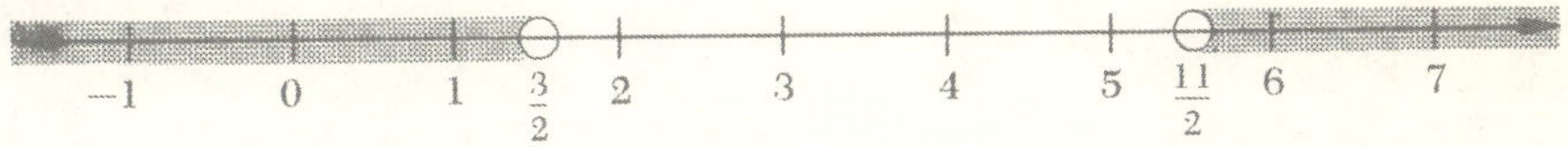

b.

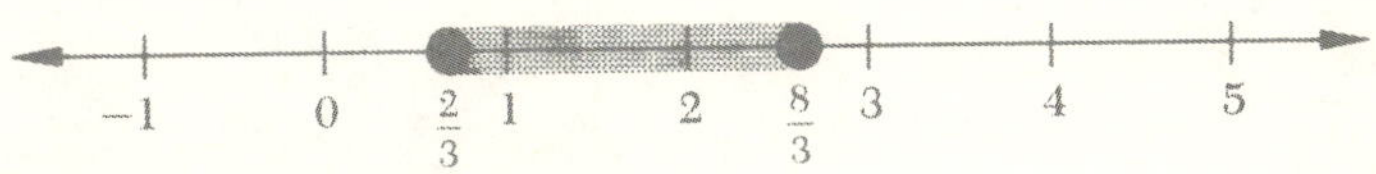

You have finished the program for Sec. 4-3. Now take the posttest to see how much you have learned.

POSTTEST 4-3

Graph the solution sets of the following open sentences on the number line:

1. $|3y - 5| = 7$
2. $|x - 3| > 5$
3. $|x + 4| \leqslant 2$
4. $|3x - 7| < 3$
5. $|4x + 5| \geqslant 6$

When you have finished the posttest, correct it yourself using the answers on page 194. Then check all the questions you got wrong on the list below:

1. ______ 4. ______

2. ______ 5. ______

3. ______

If you compare your performance with that on Pretest 4-3, you will see how much you have improved. If you made more than one mistake on this posttest, review the material and take the pretest again.

Now proceed to Final Test 4.

FINAL TEST 4

Give the solution sets of the following open sentences:

1. $|y + 3| = 6$
2. $|k| = 4$
3. $|k - 3| = -2$

4. $|5y + 3| = 7$

5. $|2y - 1| = 5$

Graph the solution sets of the following open sentences:

6. $|y - 2| = 3$

7. $|3x + 7| \leqslant 2$

8. $|4z - 1| > 3$

The answers to this test are on page 194. Check all the questions you got wrong on the list below and compare your results with those on Diagnostic Test 4 to see how much you have improved.

1. ______ 4. ______ 6. ______

2. ______ 5. ______ 7. ______

3. ______ 8. ______

If you made more than one mistake, review the material and then take the diagnostic test again. Make sure that you can get all the questions right before you finish the chapter. Now proceed to Chap. 5.

ANSWERS TO TESTS FOR CHAPTER 4

DIAGNOSTIC TEST 4

1. $\{-6, 2\}$

2. $\{-8, 8\}$

3. $\emptyset$

4. $\{-\frac{11}{3}, -1\}$

5. $\{-\frac{2}{5}, \frac{4}{5}\}$

6. 

7. 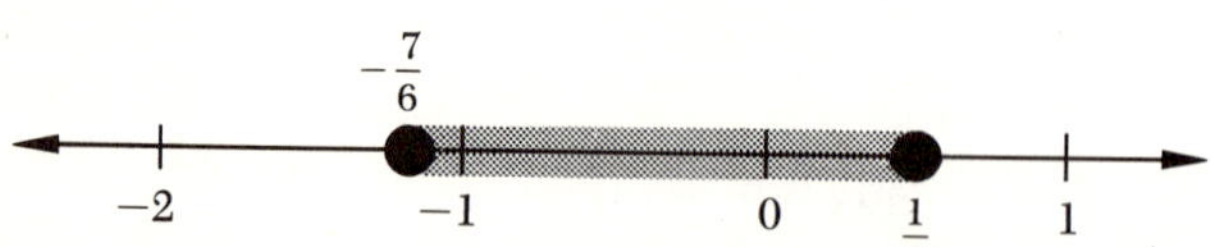

8. 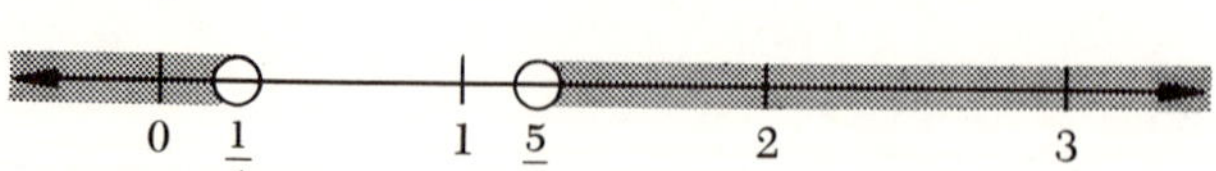

PRETEST 4-1

1. $\{-5, 9\}$
2. $\{-8, -2\}$
3. $\{-6, 6\}$
4. $\{-3\}$
5. $\emptyset$

POSTTEST 4-1

1. $\{-1, 7\}$
2. $\{-7, 3\}$
3. $\{-7, 7\}$
4. $\{-6\}$
5. $\emptyset$

PRETEST 4-2

1. $3k - 5 = 6$ and $3k - 5 = -6$
2. $\{-5, 2\}$
3. $\{\frac{4}{3}, 2\}$
4. $\{-\frac{9}{4}, -\frac{3}{4}\}$
5. $\{-\frac{2}{5}, \frac{6}{5}\}$

POSTTEST 4-2

1. $5k + 3 = 7$ and $5k + 3 = -7$
2. $\{2, 3\}$
3. $\{-\frac{8}{3}, \frac{4}{3}\}$
4. $\{-\frac{1}{2}, 2\}$
5. $\{-\frac{11}{5}, -\frac{3}{5}\}$

PRETEST 4-3

1. 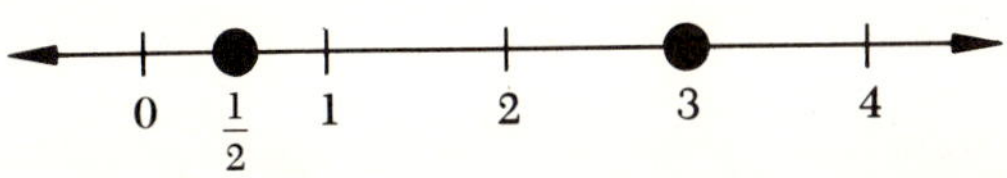

2. 

3. 

4.

5.

POSTTEST 4-3

1. 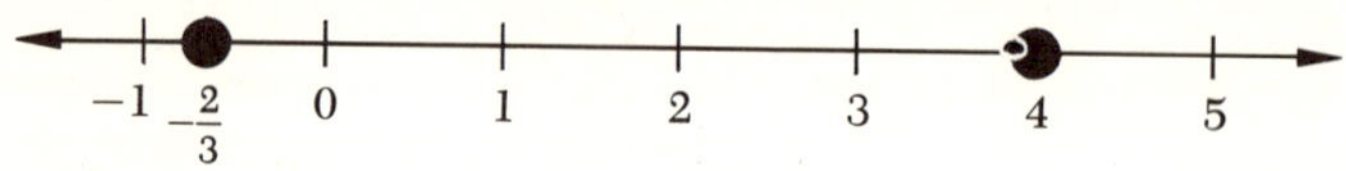

2. 

3. 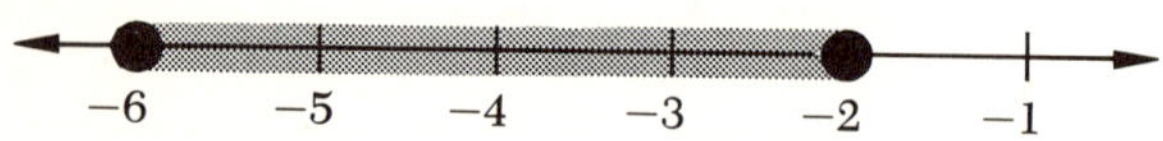

4. 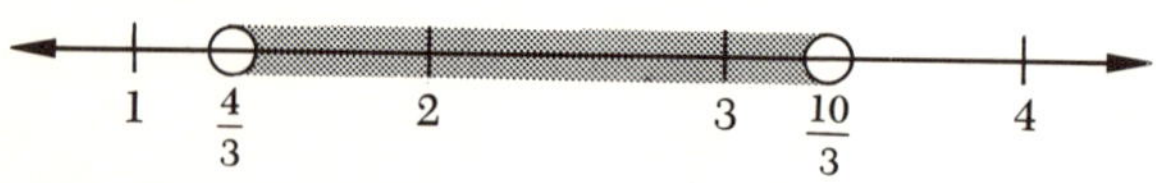

5. 

FINAL TEST 4

1. $\{-9, 3\}$
2. $\{-4, 4\}$
3. $\emptyset$
4. $\{-2, \frac{4}{5}\}$
5. $\{-2, 3\}$
6. 

7. 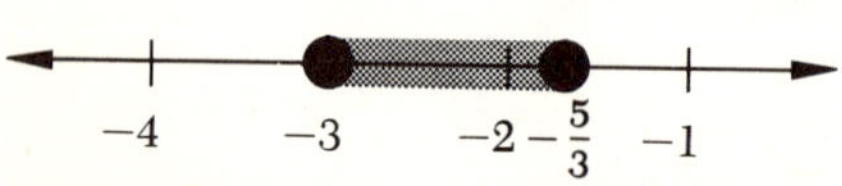

8. 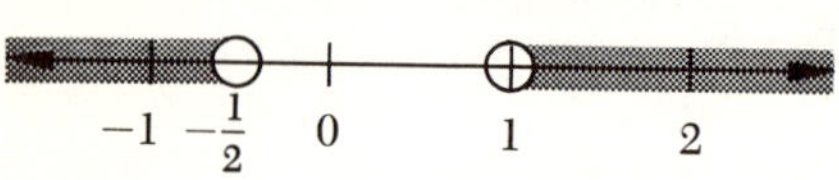

Chapter 5
First-degree Equations in Two Variables

In this chapter you will learn how to graph first-degree equations in two variables. Their graph is always an infinite straight line, which can also be defined by its slope and its Y intercept. Section 5-1 shows you how solutions of first-degree equations in two variables can be represented by ordered pairs, and Sec. 5-2 shows you how these ordered pairs can be graphed on a rectangular coordinate system. In Sec. 5-3 you learn how to construct a table of values and how equations of the form $ax + by = c$, with a and $b \neq 0$, can be graphed. Section 5-4 demonstrates how these equations can usually be converted to the form $y = mx + c$, where m is the slope and $(0, c)$ is the Y intercept. You then learn the slope-intercept method of graphing these equations. Section 5-5 shows how $y(0)$ and $y(1)$, the values of y at $x = 0$ and $x = 1$, can be found and used to determine m and graph an equation. It also discusses those cases of $ax + by = c$ for which a or b is equal to 0.

This chapter requires knowledge of Chap. 2. Throughout this chapter the domain of each variable is assumed to be the set of real numbers.

DIAGNOSTIC TEST 5

Complete the following sentences:

1. $(h, k) = (\_\_\_\_\_, -\frac{7}{4})$ is a solution of $4h + 3k = -6$.
2. $(r, s) = (-4, \_\_\_\_\_)$ is a solution of $2r - 5s = -3$.

Draw a rectangular coordinate system and graph the following ordered pairs:

3. $\left(\frac{4}{3}, -5\right)$
4. $\left(-\frac{3}{2}, -\frac{5}{4}\right)$
5. Complete the table of values and graph the equation $5x + 4y = -3$.

x	0	1	2	3	4	5
y						

6. Complete the table of values and graph the equation $3x - 5y = 5$.

x	0	1	2	3	4	5
y						

7. Graph $4x - 3y = -2$ by the slope-intercept method.

8. Complete the table of values and graph the equation from which it is derived, assuming that it is a straight line.

x	0	1	2	3	4	5
y	$-\frac{9}{4}$	$-\frac{3}{2}$				

9. Give the values of m (the slope), $y(0)$, and $y(1)$ for the equation $5x + 2y = -7$.

10. Graph the equations $x = -3$ and $y = 5$.

When you have finished the diagnostic test, correct it yourself using the answers on page 272. Then check all the questions you got wrong on the list below:

1. ______	3. ______	5. ______	7. ______
2. ______	4. ______	6. ______	8. ______
			9. ______
			10. ______

If you made no mistakes at all, you need not do any work in this chapter and should go straight on to Chap. 6, Pairs of First-degree Equations in Two Variables, which begins on page 285. But if you made any mistakes (even if it was only *one* mistake), follow these instructions:

	section	page
If you made any mistakes in Question 1 or 2, circle Sec. 5-1.	5-1	197
If you made any mistakes in Question 3 or 4, circle Sec. 5-2.	5-2	207
If you made any mistakes in Question 5 or 6, circle Sec. 5-3.	5-3	222
If you made any mistakes in Questions 7 to 10, circle Secs. 5-4 and 5-5.	5-4 5-5	233 255

You should now work through the sections whose numbers you have circled, taking them in numerical order. The sections begin on the pages indicated.

Section 5-1
Ordered-pair Solutions

Do this section if you made a mistake in Question 1 or 2 of Diagnostic Test 5. Begin by taking the pretest below, which will show you how much of Sec. 5-1 you already know and what you will be able to do by the time you have finished it.

PRETEST 5-1

Complete the following statements:

1. $(p, q) = (-3, 4)$ is a solution of $2p + 5q =$ ______.
2. $(a, b) = (-\frac{5}{3}, -\frac{2}{3})$ is a solution of $3a - 2b =$ ______.
3. $(f, g) = (-2,$ ______$)$ is a solution of $7f + 2g = -5$.
4. $(x, y) = ($______$, -\frac{3}{2})$ is a solution of $3x - 9y = -2$.

When you have finished the pretest, correct it yourself using the answers on page 276. Then check all the questions you got wrong on the list below:

1. ______ 3. ______

2. ______ 4. ______

If you made any mistakes (even if it was only *one* mistake), read the introduction and go through the program.

It you made no mistakes at all, skip Sec. 5-1, because you already know its material, and go on to the next section whose number you circled on page 196. If no further numbers were circled, take Final Test 5 (page 271).

Introduction

You will learn that the truth or falsity of an equation of the first degree in two variables depends on the values of both the variables. A solution of such an equation is normally written as an ordered pair, that is $(x,y) = (2,3)$. You will learn how to check whether a particular ordered pair is a solution or not and how to find one member of an ordered-pair solution when the other is given.

FRAMES FOR SECTION 5-1

1 The only value of x which makes $x + 3 = 5$ a true statement is ______. The solution set of the open sentence $x + 3 = 5$ is ______.

---

2 {2}

---

2 If we have an open sentence which contains two variables, its truth or falsity will normally depend on the values of both the variables. Consider, for example, the open sentence $x + y = 4$. If we put $y = 3$, it becomes $x + 3 = 4$. Is $x + 3 = 4$:

a. A true statement?
b. A false statement?
c. An open sentence whose truth or falsity depends on the value of x?

---

c. It is true when $x = 1$ but false for other values of x.

---

3 Since the open sentence $x + y = 4$ becomes a true statement when $x = 1$ and $y = 3$, we call $(x = 1, y = 3)$ a solution of $x + y = 4$.

a. Is $(x = 1, y = 2)$ a solution of $x + y = 4$?
b. Is $(x = 2, y = 2)$ a solution of $x + y = 4$?

---

a. no *b.* yes

---

4 Suppose we have the open sentence $x + y = 8$.

a. If we put $x = 3$, what value of y will make the sentence a true statement?
b. $(x = 3, y =$ ______$)$ is a solution of $x + y = 8$.
c. Could you choose any other value of y to go with $x = 3$ as a solution of $x + y = 8$?

---

a. 5 *b.* 5 *c.* no

---

5 $(x = 7, y =$ ______$)$ is another solution of $x + y = 8$. Could you choose any other value of y to go with $x = 7$ as a solution of $x + y = 8$?

1; no

---

6 Instead of writing ($x = 7$, $y = 1$) we write $(x,y) = (7,1)$. This shows that the first number in each pair is the value of x and the second number is the value of y.

a. $(x,y) = (3,5)$ means that $x =$ ______ and $y =$ ______.
b. $(x,y) = (2,6)$ means that $x =$ ______ and $y =$ ______.

---

a. 3; 5 *b.* 2; 6

---

7 (7,1), (3,5), (2,6), etc., are called *ordered pairs* because the order of the two numbers determines what they represent. Thus

a. $(x,y) = (7,1)$ means that $x =$ ______ and $y =$ ______.
b. $(x,y) = (1,7)$ means that $x =$ ______ and $y =$ ______.

---

a. 7; 1 *b.* 1; 7

---

8 *a.* Is $(x,y) = (3,2)$ a solution of $x + 2y = 7$?
b. Is $(x,y) = (2,3)$ a solution of $x + 2y = 7$?
c. Is $(x,y) = (3,1)$ a solution of $x + 2y = 7$?
d. Is $(x,y) = (1,3)$ a solution of $x + 2y = 7$?

---

a. yes *b.* no *c.* no *d.* yes

---

9 *a.* To complete the statement

$(h,k) = (1,2)$ is a solution of $3h - 2k = ?$

we substitute $h =$ ______, $k =$ ______ to get

$3h - 2k = 3(1) - 2(2) = 3 - 4 = -1$

Hence $(h,k) = (1,2)$ is a solution of $3h - 2k =$ ______.

b. To complete the statement

$(h,k) = (-2,3)$ is a solution of $2h - 3k = ?$

we substitute $h =$ ______, $k =$ ______ to get

$2h - 3k =$ ______ $-$ ______ $=$ ______ $-$ ______ $=$ ______

Hence $(h,k) = (-2,3)$ is a solution of $2h - 3k =$ ______.

---

a. $1; 2; -1$ *b.* $-2; 3; 2(-2) - 3(3) = -4 - 9 = -13; -13$

---

10 *a.* To complete the statement

$(a,b) = (-4,-5)$ is a solution of $2a - 3b = ?$

we substitute $a =$ ______, $b =$ ______ to get

$2a - 3b =$ ______ $-$ ______ $=$ ______ $+$ ______ $=$ ______

Hence $(a,b) = (-4,-5)$ is a solution of $2a - 3b =$ ______.

b. Complete the statement:

$(a,b) = (2,-3)$ is a solution of $4a + 5b =$ ______.

(Show your work. Don't try to do it in your head.)

---

a. $-4; -5; 2(-4) - 3(-5) = -8 + 15 = 7; 7$

b. Substituting $a = 2$, $b = -3$, we get

$$4(2) + 5(-3) = 8 + (-15) = -7$$
$$4a + 5b = -7$$

---

11 Complete the statements:

a. $(m,n) = (-2,3)$ is a solution of $2m + 3n =$ ________________.

b. $(m,n) = (-2,-3)$ is a solution of $2m + 3n =$ ________________.

c. $(m,n) = (2,-\frac{1}{3})$ is a solution of $2m + 3n =$ ________________.

Show your work.

---

a. $2(-2) + 3(3) = -4 + 9 = 5$

b. $2(-2) + 3(-3) = -4 + (-9) = -13$

c. $2(2) + 3(-\frac{1}{3}) = 4 + (-1) = 3$

---

12 Complete the statements:

a. $(c,d) = (-2,5)$ is a solution of $3c - 4d =$ ________________.

b. $(c,d) = (5,-2)$ is a solution of $3c - 4d =$ ________________.

c. $(c,d) = (-2,-\frac{3}{4})$ is a solution of $3c - 4d =$ ________________.

Show your work.

---

a. $3(-2) - 4(5) = -6 - 20 = -26$

b. $3(5) - 4(-2) = 15 + 8 = 23$

c. $3(-2) - 4(-\frac{3}{4}) = -6 + 3 = -3$

---

13 Use exactly the same procedure in checking solutions.

a. To check if $(f,g) = (-3,4)$ is a solution of $2f - 3g = -18$, substitute $f =$ ______, $g =$ ______. Thus

$2f - 3g = 2(-3) - 3(4) =$ ______ $-$ ______ $=$ ______

Since the answer does come to -18, you know that $(f,g) = ($______, ______$)$ is a solution of $2f - 3g = -18$.

b. To check if $(a,b) = (4,-2)$ is a solution of $3a - 4b = 21$, substitute $a =$ ______, $b =$ ______. Thus

$3a - 4b =$ ______ $-$ ______ $=$ ______ $+$ ______ $=$ ______

Since the answer comes to ______, you know that $(a,b) = ($______, ______$)$ is *not* a solution of $3a - 4b = 21$.

---

a. -3; 4; $-6 - 12 = -18$; $(-3,4)$
b. 4; -2; $3(4) - 4(-2) = 12 + 8 = 20$; 20; $(4,-2)$

---

14 *a.* Is $(m,n) = (-2,-3)$ a solution of $3m + 2n = -10$?

b. Is $(h,k) = \left(\frac{2}{3},-5\right)$ a solution of $3h - k = -3$?

c. Is $(y,z) = \left(\frac{3}{4},-\frac{2}{3}\right)$ a solution of $8y + 3z = 4$?

Show your work.

---

a. $3(-2) + 2(-3) = -6 + (-6) = -12$; no

b. $3\left(\frac{2}{3}\right) - (-5) = 2 + 5 = 7$; no

c. $8\left(\frac{3}{4}\right) + 3\left(-\frac{2}{3}\right) = 6 - 2 = 4$; yes

---

15 *a.* Is $(a,b) = (-1,5)$ a solution of $3a - 2b = -7$?

b. Is $(c,d) = \left(\frac{1}{3},\frac{2}{3}\right)$ a solution of $5c + 2d = 3$?

c. Is $(e,f) = \left(-\frac{1}{4},-2\right)$ a solution of $28e - 4f = -1$?

Show your work.

---

a. $3(-1) - 2(5) = -3 - 10 = -13$; no

b. $5\left(\frac{1}{3}\right) + 2\left(\frac{2}{3}\right) = \frac{5}{3} + \frac{4}{3} = \frac{9}{3} = 3$; yes

c. $28\left(-\frac{1}{4}\right) - 4(-2) = -7 + 8 = 1$; no

---

16 To complete the statement $(x,y) = (2,?)$ is a solution of $3x + 5y = 8$ we substitute $x =$ ______ to get the open sentence $3(2) + 5y = 8 \longleftrightarrow 6 + 5y = 8$, which we can solve for y.

$6 + 5y = 8 \longleftrightarrow 5y =$ ______ $\longleftrightarrow y =$ ______

Hence $(x,y) = (2,\_\_)$ is a solution of $3x + 5y = 8$.

CHECK

$3(2) + 5(\_\_) =$ ______ $+$ ______ $=$ ______

---

2 2 $\frac{2}{5}$ $\frac{2}{5}$

CHECK

$3(2) + 5\left(\frac{2}{5}\right) = 6 + 2 = 8$

---

17 To complete the statement $(x,y) = (?,3)$ is a solution of $3x + 5y = 8$, we substitute $y =$ ______ to get the open sentence $3x + 5(3) = 8 \longleftrightarrow 3x + 15 = 8$ which we can solve for x.

$3x + 15 = 8 \longleftrightarrow 3x =$ ______ $\longleftrightarrow x =$ ______

Hence $(x,y) =$ (______, 3) is a solution of $3x + 5y = 8$.

CHECK

3(______) + 5 (______) = ______ + ______ = ______

$3 \quad -7 \quad -\frac{7}{3} \quad -\frac{7}{3}$

CHECK

$3\left(-\frac{7}{3}\right) + 5(3) = (-7) + 15 = 8$

---

18 *a.* To complete the statement

$(x,y) = (-1,?)$ is a solution of $2x - 3y = 9$

we substitute $x =$ ______ to get the open sentence

$2(\_\_\_\_) - 3y = 9 \longleftrightarrow \_\_\_\_ - 3y = 9 \longleftrightarrow -3y = \_\_\_\_$

$\longleftrightarrow y = \_\_\_\_$

Hence $(-1,\_\_)$ is a solution of $2x - 3y = 9$. Check this answer.

b. Complete the statement $(x,y) = (\_\_,2)$ is a solution of $2x - 3y = 9$. Show all your work and check your answer.

---

a. $-1; -1; -2; 11; -\frac{11}{3}; -\frac{11}{3}$

CHECK

$2(-1) - 3(-\frac{11}{3}) = -2 + 11 = 9$

b. Substituting $y = 2$, we get

$$2x - 3(2) = 9$$
$$2x - 6 = 9$$
$$2x = 15$$
$$x = \frac{15}{2}$$
$$(x,y) = \left(\frac{15}{2},2\right)$$

is a solution of $2x - 3y = 9$

CHECK

$2\left(\frac{15}{2}\right) - 3(2) = 15 - 6 = 9$

---

19 Complete the statements:

a. $(m, n) = (-2, \_\_)$ is a solution of $3m - 4n = 8$.

b. $(h, k) = \left(\_\_, -\frac{1}{3}\right)$ is a solution of $5h + 6k = -7$.

Show all your work and check your answers.

---

a. Substituting $m = -2$, we get

$$3(-2) - 4n = 8$$
$$-6 - 4n = 8$$
$$-4n = 14$$
$$n = -\frac{7}{2}$$

Hence $(m, n) = \left(-2, -\frac{7}{2}\right)$ is a solution of $3m - 4n = 8$.

CHECK

$$3(-2) - 4\left(-\frac{7}{2}\right) = -6 + 14 = 8$$

b. Substituting $k = -\frac{1}{3}$, we get

$$5h + 6\left(-\frac{1}{3}\right) = -7$$
$$5h + (-2) = -7$$
$$5h = -5$$
$$h = -1$$

Hence $(h, k) = \left(-1, -\frac{1}{3}\right)$ is a solution of $5h + 6k = -7$.

CHECK

$$5(-1) + 6\left(-\frac{1}{3}\right) = -5 + (-2) = -7$$

---

20 Complete the statements:

a. $(f, g) = (-3, \_\_\_\_)$ is a solution of $4f - 3g = 6$.

b. $(c, d) = \left(\_\_\_\_, -\frac{3}{4}\right)$ is a solution of $3c + 12d = 8$.

Show all your work and check your answers.

a. Substituting $f = -3$, we get

$$\begin{aligned} 4(-3) - 3g &= 6 \\ -12 - 3g &= 6 \\ -3g &= 18 \\ g &= -6 \end{aligned}$$

Hence $(f, g) = (-3, -6)$ is a solution of $4f - 3g = 6$.

CHECK

$$4(-3) - 3(-6) = -12 + 18 = 6$$

b. Substituting $d = -\frac{3}{4}$, we get

$$\begin{aligned} 3c + 12\left(-\frac{3}{4}\right) &= 8 \\ 3c + (-9) &= 8 \\ 3c &= 17 \\ c &= \frac{17}{3} \end{aligned}$$

Hence $(c, d) = \left(\frac{17}{3}, -\frac{3}{4}\right)$ is a solution of $3c + 12d = 8$.

CHECK

$$3\left(\frac{17}{3}\right) + 12\left(-\frac{3}{4}\right) = 17 + (-9) = 8$$

You have finished the program for Sec. 5-1. Now take the posttest to see how much you have learned.

POSTTEST 5-1

Complete the following statements:

1. $(m,n) = (-5,3)$ is a solution of $5m + 2n =$ ______.
2. $(c, d) = (-\frac{3}{4}, -\frac{5}{4})$ is a solution of $7c - 2d =$ ______.
3. $(h, k) = (-3,$ ______$)$ is a solution of $2h + 5k = -7$.
4. $(r, s) = ($______$, -\frac{5}{3})$ is a solution of $3r - 4s = -4$.

When you have finished the posttest, correct it yourself using the answers on page 276. Then check all the questions you got wrong on the list below:

1. ______	3. ______
2. ______	4. ______

If you compare your performance with that on Pretest 5-1, you will see how much you have improved. If you made more than one mistake on this posttest, review the material and take the pretest again. Don't move on to the next section until you can get all the questions right.

Now proceed to the next section whose number you circled on page 196. If no other sections were circled, take Final Test 5 (page 271).

Section 5-2
The Rectangular Coordinate System

Do this section if you made a mistake in Question 3 or 4 of Diagnostic Test 5. Begin by taking the pretest below, which will show you how much of Sec. 5-2 you already know and what you will be able to do by the time you have finished it.

PRETEST 5-2

Draw a rectangular coordinate system and graph the following ordered pairs:

1. $(\frac{11}{5}, -\frac{3}{2})$
2. $(-\frac{7}{3}, -\frac{5}{3})$
3. $(0, -\frac{9}{2})$
4. $(-\frac{7}{4}, \frac{3}{2})$

When you have finished the pretest, correct it yourself using the answers on page 276. Then check all the questions you got wrong on the list below:

1. ______ 3. ______

2. ______ 4. ______

If you made any mistakes (even if it was only *one* mistake), read the introduction and go through the program.

If you made no mistakes at all, skip Sec. 5-2, because you already know its material, and go on to the next section whose number you circled on page 196. If no further numbers were circled, take Final Test 5 (page 271).

Introduction

After learning about the rectangular coordinate system with X and Y axes as a two-dimensional version of the number line, you are shown how ordered pairs are in one-to-one correspondence with points on the coordinate system and how they can be graphed on it.

FRAMES FOR SECTION 5-2

1 Graph the solution of $x + 2 = 4$ on a number line.

---

2 *a.* $(x, y) = (1, \_\_\_\_)$ is a solution of $x + 2y = 5$.
b. $(x, y) = (\_\_\_\_, 3)$ is a solution of $x + 2y = 5$.
c. Could you graph these solutions on the number line?

---

a. 2 *b.* -1 *c.* no

---

3 The solution set of an equation in *one* variable is graphed in *one* dimension, i.e., on a line. But the solution set of an equation in *two* variables has to be graphed in ____ dimensions, i.e., on a plane.

---

two

---

4 The usual method of graphing an ordered pair in a plane uses the rectangular coordinate system, as shown:

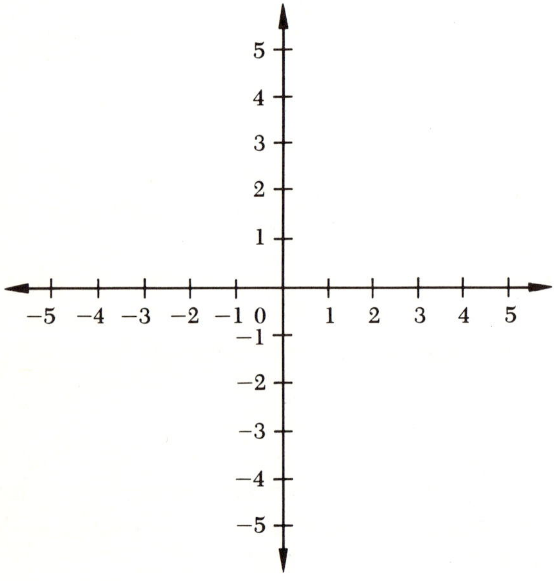

This system consists of two ordered number lines (called *axes*) which form an angle of ____° with each other and which intersect at the point ____ on each line. The positive end of the horizontal axis is on the (left/right) ____ and the positive end of the vertical axis is on the (top/bottom) ____.
Note: Horizontal goes from side to side like the *horizon*.

90°; 0; right; top

5

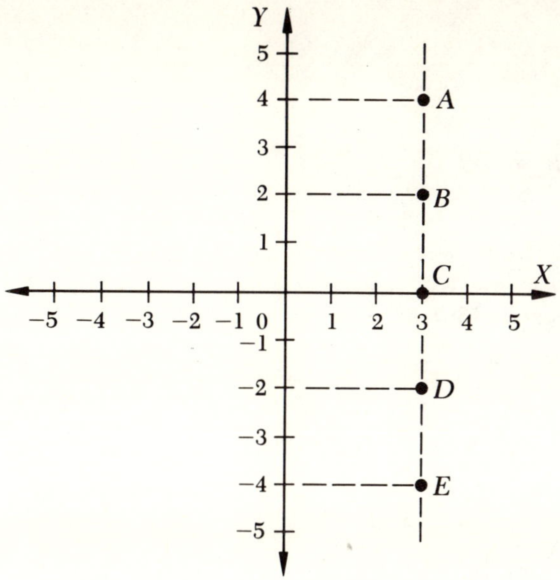

Point A represents the ordered pair $(x,y) = (3,4)$.
Point B represents the ordered pair $(x,y) = (3,2)$.

a. Point C represents the ordered pair $(x,y) = (\_,0)$.
b. Point D represents the ordered pair $(x,y) = (\_,\_)$.
c. Point E represents the ordered pair $(x,y) = (\_,\_)$.

---

a. 3 *b.* (3,−2) *c.* (3,−4)

6

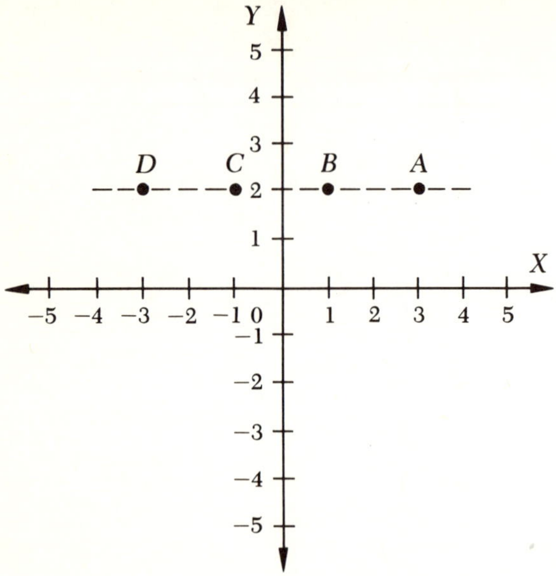

Point *A* represents the ordered pair (3,2).

a. Point *B* represents the ordered pair (_,2).
b. Point *C* represents the ordered pair (_,_).
c. Point *D* represents the ordered pair (_,_).

---

a. 1 *b.* (−1,2) *c.* (−3,2)

---

7 Which of the following ordered pairs are represented by points on the dashed line in Frame 5? Which pairs are represented by points on the dashed line in Frame 6?

a. (3,−1)
b. (−1,3)
c. (3,2)
d. (2,3)

---

a and *c* are on the dashed line in Frame 5. Only *c* is on the dashed line in Frame 6.

---

8 Usually a grid is drawn to make it easier to locate a point accurately.

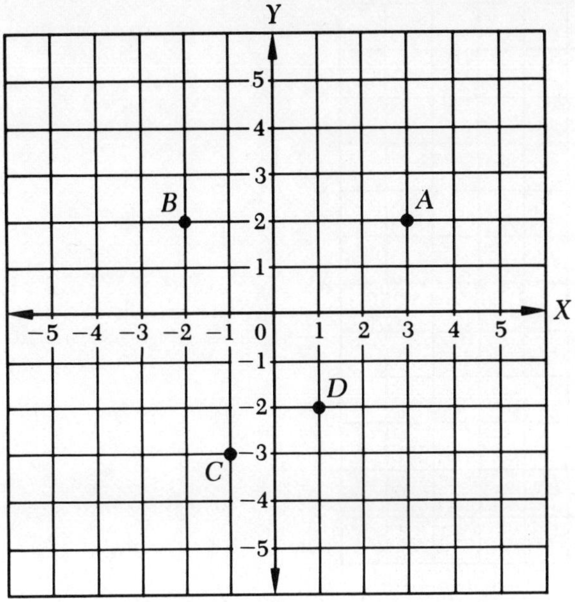

a. Does point *A* represent the ordered pair $(x,y) = (3,2)$ or $(x,y) = (2,3)$?

b. Does point *B* represent the ordered pair $(x,y) = (2,-2)$ or $(x,y) = (-2,2)$?

c. Does point *C* represent the ordered pair $(x,y) = (-1,-3)$ or $(x,y) = (-3,-1)$?

d. Does point *D* represent the ordered pair $(x,y) = (1,-2)$ or $(x,y) = (-2,1)$?

---

a. (3,2) *b.* (−2,2) *c.* (−1,−3) *d.* (1,−2)

9

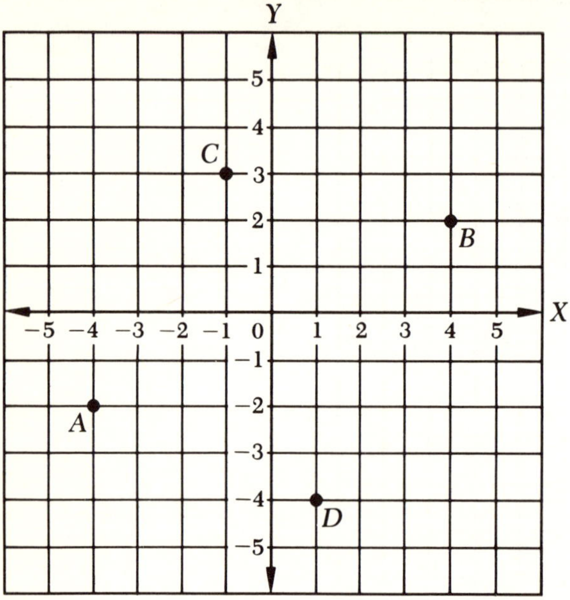

a. Does point A represent the ordered pair $(x,y) = (-4,-2)$ or $(x,y) = (-2,-4)$?

b. Point B represents the ordered pair $(x,y) = (\_,\_)$.

c. Point C represents the ordered pair $(x,y) = (\_,\_)$.

d. Point D represents the ordered pair ______________.

---

a. $(-4,-2)$ *b.* $(4,2)$ *c.* $(-1,3)$ *d.* $(x,y) = (1,-4)$

10

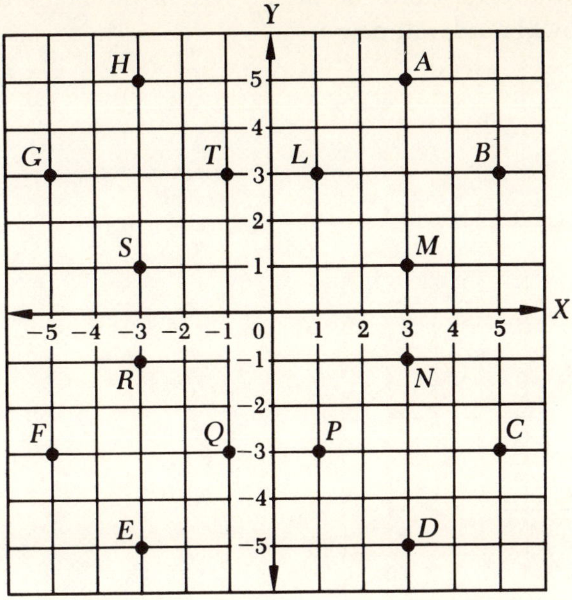

a. Which point represents the ordered pair $(x,y) = (-5,3)$?
b. Which point represents the ordered pair $(x,y) = (-3,-5)$?
c. Which point represents the ordered pair $(x,y) = (1,3)$?
d. Which point represents the ordered pair $(x,y) = (3,-1)$?

---

a. G *b.* E *c.* L *d.* N

---

11 Use graph paper to draw a rectangular coordinate system and graph the points representing the following ordered pairs:

$A = (x,y) = (5,4)$
$B = (x,y) = (-3,2)$
$C = (x,y) = (-1,-5)$
$D = (x,y) = (4,-1)$

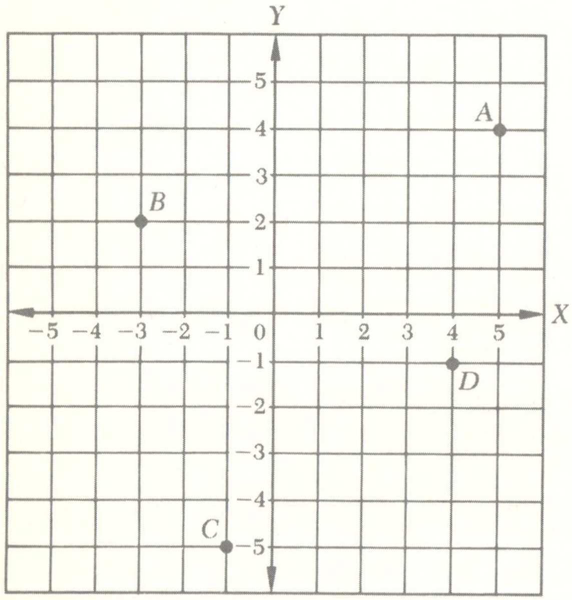

---

12 *a.* At every point on the X axis (the horizontal number line), the value of y is ______.

b. At every point on the Y axis (the vertical number line), the value of x is ______.

c. The *origin* (the point where the two axes intersect) represents the ordered pair (_,_).

---

a. 0 *b.* 0 *c.* (0,0)

---

13 Which of the following pairs are represented by points on the y axis? Which ot the following pairs are represented by points on the x axis?

a. $(x,y) = (0,4)$
b. $(x,y) = (4,0)$
c. $(x,y) = (0,-1)$
d. $(x,y) = (-1,0)$
e. $(x,y) = (0,0)$

---

a, *c*, and *e* are on the *Y* axis; *b*, *d*, and *e* are on the *X* axis.

14

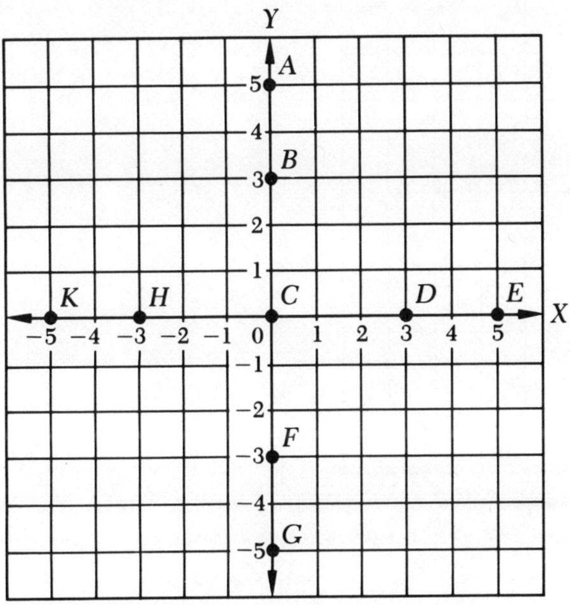

a. Which point represents the ordered pair $(x,y) = (3,0)$?
b. Which point represents the ordered pair $(x,y) = (0,-3)$?
c. Which point represents the ordered pair $(x,y) = (0,5)$?
d. Which point represents the ordered pair $(x.y) = (-5,0)$?
e. Which point represents the ordered pair $(x.y) = (0,0)$?

---

a. *D* *b.* *F* *c.* *A* *d.* *K* *e.* *C*

---

15 Draw a rectangular coordinate system and graph points representing the following ordered pairs:

$A = (x,y) = (-4,-1)$
$B = (x,y) = (-2,0)$
$C = (x,y) = (5,-2)$
$D = (x,y) = (0,1)$

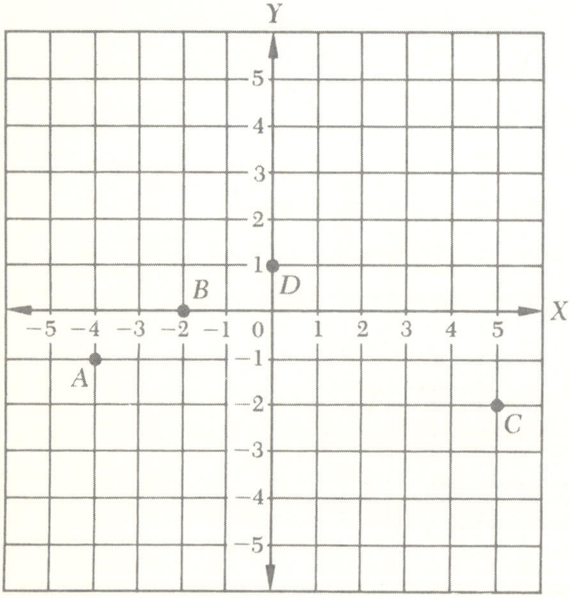

---

16 Sometimes an ordered pair such as $(x, y) = (2, 3)$ is written in abbreviated form as $(2,3)$. In such cases the first number in the pair always refers to (x/y) ______ and is graphed on the (horizontal/vertical) ____________ axis. Draw a rectangular coordinate system and graph the pairs $(-2, -4)$ and $(0, 3)$.

---

x; horizontal

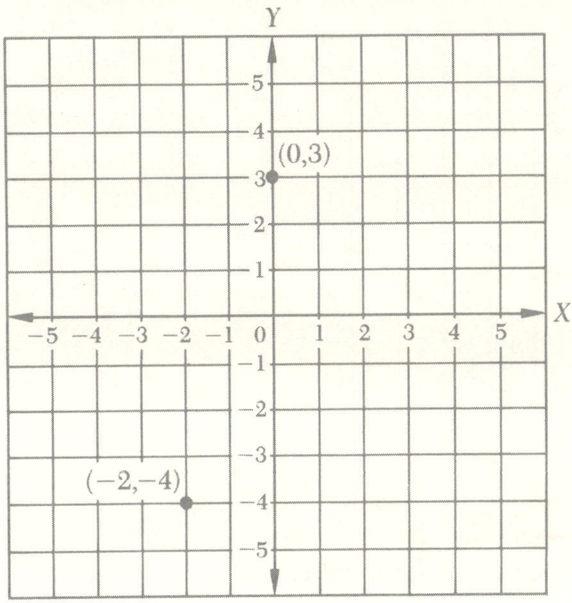

---

17 Draw a rectangular coordinate system and graph the following pairs.

$A = (3,-5)$ $B = (-2,0)$ $C = (2,-4)$ $D = (-3,-1)$

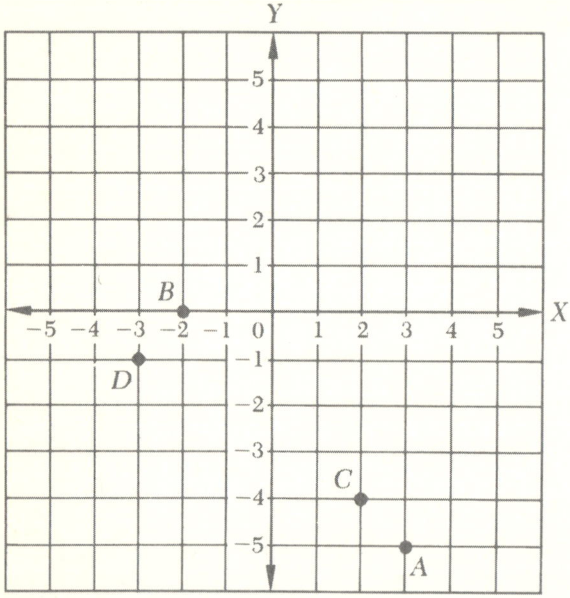

18 Graph the following on sections of the number line:

a. $x = \frac{2}{3}$

b. $x = -\frac{10}{3}$

c. $x = -\frac{7}{4}$

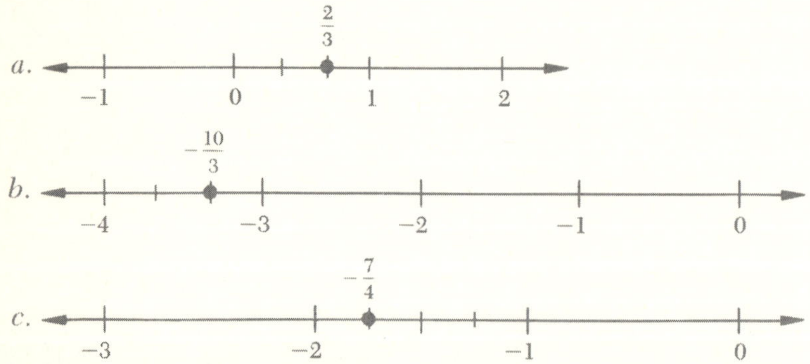

19 Use graph paper to draw a rectangular coordinate system and graph the following ordered pairs:

$A = (\frac{2}{3}, 1)$ $B = \frac{5}{2}, \frac{3}{2}$ $C = (-\frac{7}{3}, -\frac{3}{2})$ $D = (-\frac{2}{3}, -\frac{8}{3})$

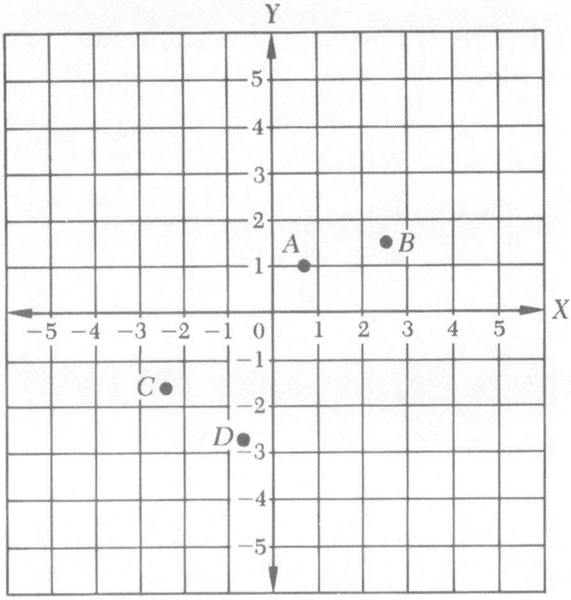

20 Use graph paper to draw a rectangular coordinate system and graph the following ordered pairs:

$A = (\frac{9}{4}, -\frac{1}{4})$ $\quad B = (0, -\frac{10}{3})$ $\quad C = (-\frac{9}{5}, \frac{11}{2})$ $\quad D = (-\frac{11}{3}, -\frac{13}{3})$

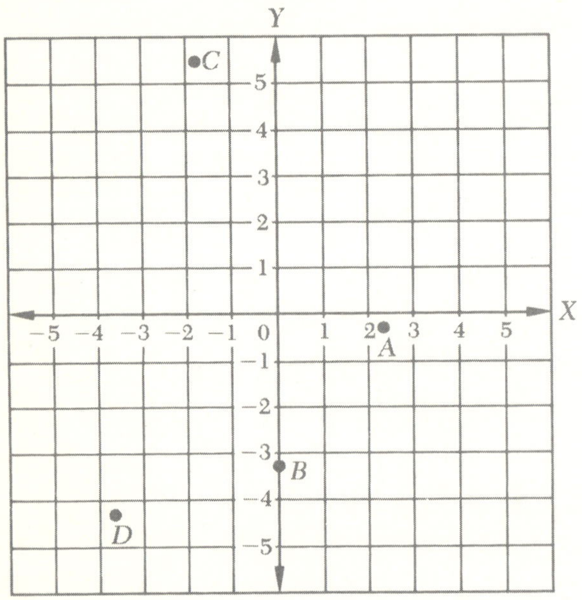

You have finished the program for Sec. 5-2. Now take the posttest to see how much you have learned.

POSTTEST 5-2

Draw a rectangular coordinate system and graph the following ordered pairs:

1. $(-\frac{7}{2}, \frac{8}{5})$
2. $(-\frac{9}{4}, -\frac{7}{4})$
3. $(-\frac{5}{2}, 0)$
4. $(\frac{11}{3}, -\frac{8}{3})$

When you have finished the posttest, correct it yourself using the answers on page 276. Then check all the questions you got wrong on the list below:

1. ______ 3. ______

2. ______ 4. ______

If you compare your performance with that on Pretest 5-2, you will see how much you have improved. If you made more than one mistake on this post-test, review the material and take the pretest again. Don't move on to the next section until you can get all the questions right.

Now proceed to the next section whose number you circled on page 196. If no other sections were circled, take Final Test 5 (page 271).

Section 5-3
Graphing from a Table of Values

Do this section if you made a mistake in Question 5 or 6 of Diagnostic Test 5. Begin by taking the pretest below, which will show you how much of Sec. 5-3 you already know, and what you will be able to do by the time you have finished it.

PRETEST 5-3

For each of the following equations complete the table of values and graph the solution sets:

1. $2x + 5y = 3$

x	-1	0	1	2	3	4
y						

2. $3x - 4y = -2$

x	-2	-1	0	1	2	3
y						

3. $5x + 2y = -1$

x	-1	0	1	2	3	4	5
y							

When you have finished the pretest, correct it yourself using the answers on page 277. Then check all the questions you got wrong on the list below:

1. ______ 3. ______

2. ______

If you made any mistakes (even if it was only *one* mistake), read the introduction and go through the program.

If you made no mistakes at all, skip Sec. 5-3, because you already know its material, and go on to the next section whose number you circled on page 196. If no further numbers were circled, take Final Test 5 (page 271).

Introduction

First you are shown that the graph of a first-degree equation in two variables is an infinite straight line, that all points on the line are solutions, and that all points off the line are not solutions. Then with the methods of Sec. 5-1 you construct a table of values for an equation and use these values to determine the position of the line. You learn how to apply the method to all equations of the form $ax + by = c$ (with a, b, and c real numbers and a and b not equal to 0).

FRAMES FOR SECTION 5-3

1 Complete the following ordered pairs so as to make them solutions of $2x + y = 2$:

a. $(x,y) = (-2,\_)$
b. $(x,y) = (-1,\_)$
c. $(x,y) = (0,\_)$
d. $(x,y) = (1,\_)$
e. $(x,y) = (2,\_)$

---

a. 6 *b.* 4 *c.* 2 *d.* 0 *e.* −2

---

2 Draw a rectangular coordinate system and graph these five ordered pairs. What do you notice about the five points you have graphed?

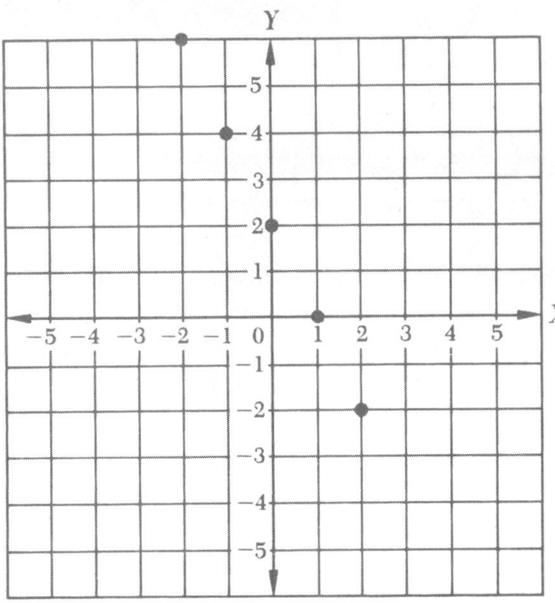

All five points are on a straight line.

---

3 Draw a line through the five points.

a. Which of the following points are on the line?

$(-\frac{3}{2},5)$ $(-1,3)$ $(\frac{1}{2},\frac{1}{2})$ $(2,-1)$ $(3,-4)$

b. Which of the above ordered pairs is a solution of $2x + y = 2$?

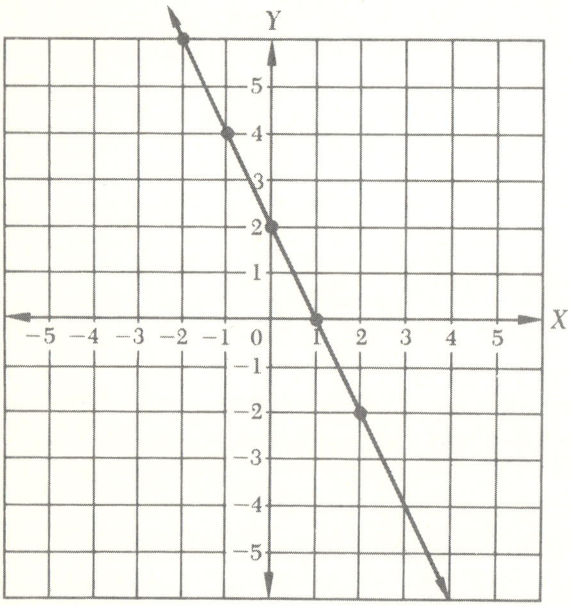

The arrows show that the line goes on forever in both directions.

a. $(-\frac{3}{2},5)$; $(3,-4)$ *b.* $(-\frac{3}{2},5)$; $(3,-4)$

---

4 There is a one-to-one correspondence between the points on the line and the solutions of $2x + y = 2$. Every ordered pair which is a solution of $2x + y = 2$ is represented by a point which is ________________. Every ordered pair which is *not* a solution of $2x + y = 2$ is represented by a point which is ________________. The line is the graph of the open sentence ________________.

---

on the line; not on the line; $2x + y = 2$

---

5 It can be shown that any open sentence of the form $ax + by = c$ (with a, b, and c real numbers and a and b not equal to 0) has as infinite straight line for its graph. If the straight line below is the graph of an open sentence of this type, which of the following ordered pairs are solutions?

$(-4,-5)$ $(4,5)$ $(-1,-1)$ $(-1,-2)$ $(-3,-4)$ $(-1,1)$

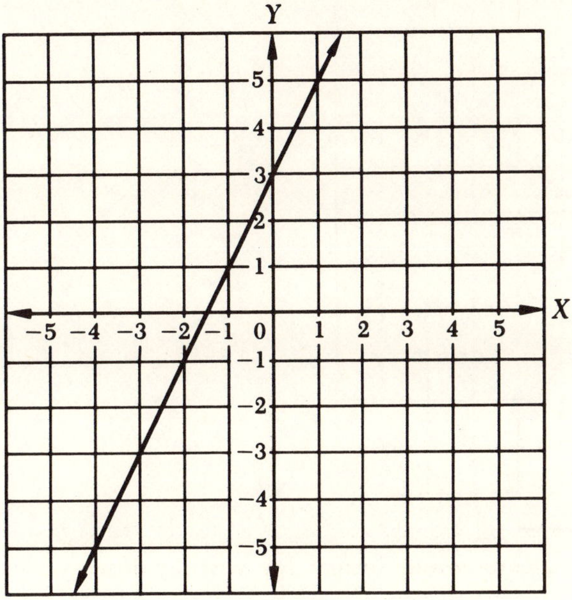

$(-4,-5)$ $(-1,1)$

6 Complete the following ordered pairs to make them solutions of $3x - 4y = 1$:

a. $(x,y) = (-5,\_)$
b. $(x,y) = (-3,\_)$
c. $(x,y) = (-1,\_)$
d. $(x,y) = (1,\_)$
e. $(x,y) = (3,\_)$

a. -4 *b.* $-\frac{5}{2}$ *c.* -1 *d.* $\frac{1}{2}$ *e.* 2

7 Use these five solutions to graph the solution set of the open sentence $3x - 4y = 1$.

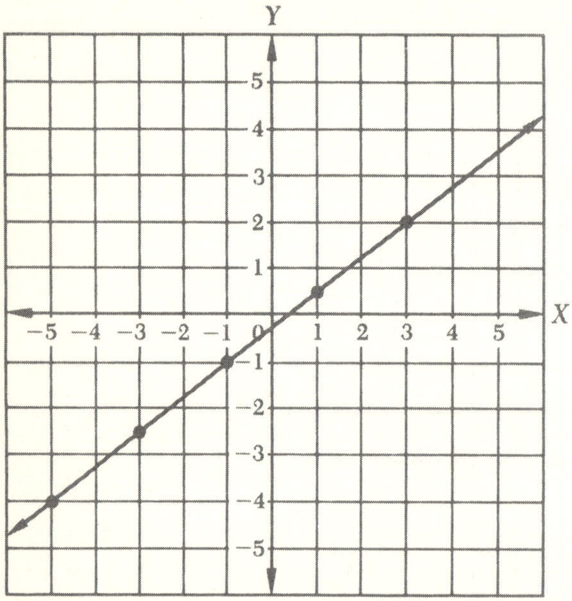

---

8 The *table of values* is a convenient format for writing down ordered pairs which are solutions of a particular open sentence. The table below, for example, indicates that the ordered pairs $(-2,7)$, $(-1,\_)$, $(\_,\_)$, $(\_,\_)$, $(\_,\_)$, $(\_,\_)$, $(\_,\_)$ are solutions of the open sentence $x + y = 5$.

x	−2	−1	0	1	2	3	4
y	7	6	5	4	3	2	1

---

6 (0,5) (1,4) (2,3) (3,2) (4,1)

---

9 The following ordered pairs are solutions of the open sentence $x + y = 3$.

(−2,5) (−1,4) (0,3) (1,2) (2,1)

Make a table of values for the open sentence $x + y = 3$.

---

x	−2	−1	0	1	2
y	5	4	3	2	1

---

10 Use this table of values to graph the solution set of $x + y = 3$.

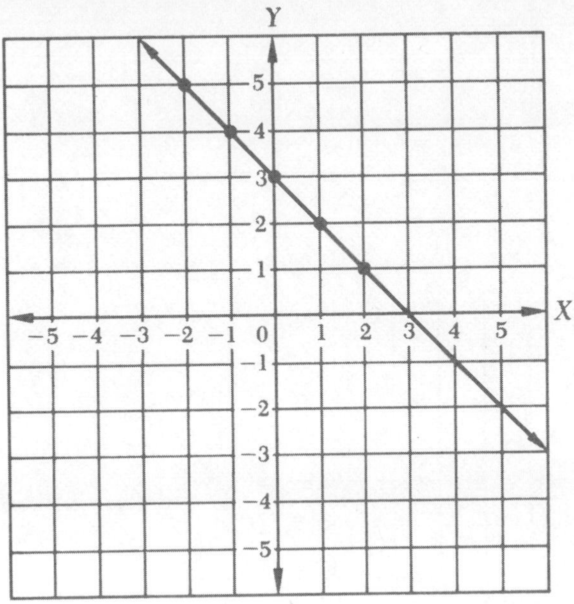

Did you remember the arrows? They show that the line is infinite.

11 Complete the table of values and graph the solution set of $x - 4y = 5$.

x	-3	-1	1	3	5
y					

---

x	-3	-1	1	3	5
y	-2	$-\frac{3}{2}$	-1	$-\frac{1}{2}$	0

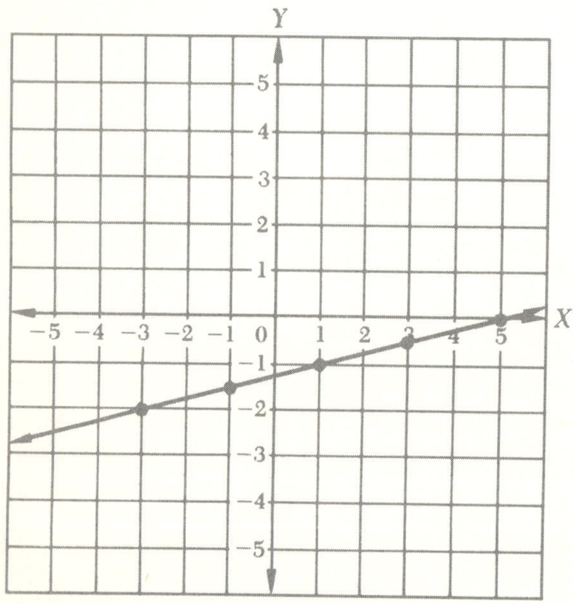

---

12 Complete the table of values and graph the solution set of $2x - 3y = -6$.

x	-3	-1	0	1	3
y					

---

x	-3	-1	0	1	3
y	0	$\frac{4}{3}$	2	$\frac{8}{3}$	4

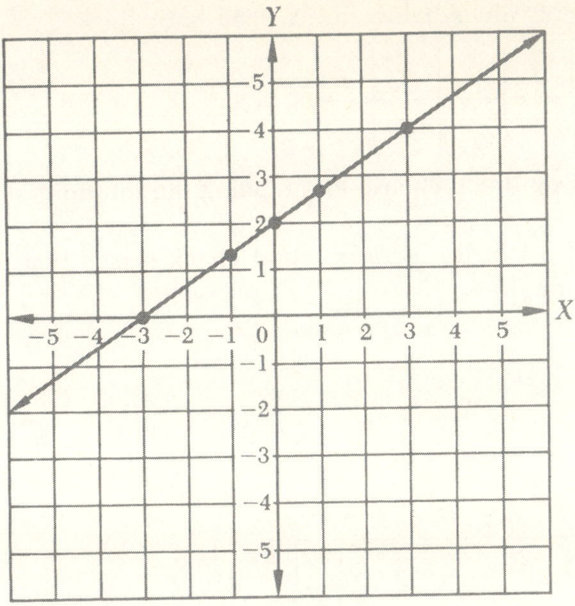

---

13 You need only two points in order to graph a straight line, but it is useful to use other points as a check. For accurate graphing, it is best to use points with *integral* values of x and y and points which are *far* apart. Thus in Frame 12 you would use points (_,_) and (_,_) for drawing the line and points (_,_), (_,_), and (_,_) as a check.

$(-3,0)$ $(3,4)$ $(-1,\frac{4}{3})$ $(0,2)$ $(1,\frac{8}{3})$

---

14 Consider the following tables of values. Which pairs of points would you choose for graphing the line?

a.

x	-3	-1	0	1	2	3
y	-2	$-\frac{2}{3}$	0	$\frac{2}{3}$	$\frac{4}{3}$	2

b.

x	-1	0	1	2	3	5
y	3	$\frac{7}{3}$	$\frac{5}{3}$	1	$\frac{1}{3}$	-1

c.

x	0	1	2	3	4
y	$\frac{7}{3}$	3	$\frac{11}{3}$	$\frac{13}{3}$	5

---

a. $(-3,-2)$; $(3,2)$ *b.* $(-1,3)$; $(5,-1)$ *c.* $(1,3)$; $(4,5)$

---

15 *a.* Complete the table of values below for $3x - 5y = 8$:

x	-4	-2	0	2	4	6

b. Which two points would you choose for graphing the solution set of $3x - 5y = 8$?

c. Graph the solution set of $3x - 5y = 8$ and check to see that the other points are on the line.

---

a.

x	-4	-2	0	2	4	6
y	-4	$-\frac{14}{5}$	$-\frac{8}{5}$	$-\frac{2}{5}$	$\frac{4}{5}$	2

b. $(-4,-4)$; $(6,2)$

c.

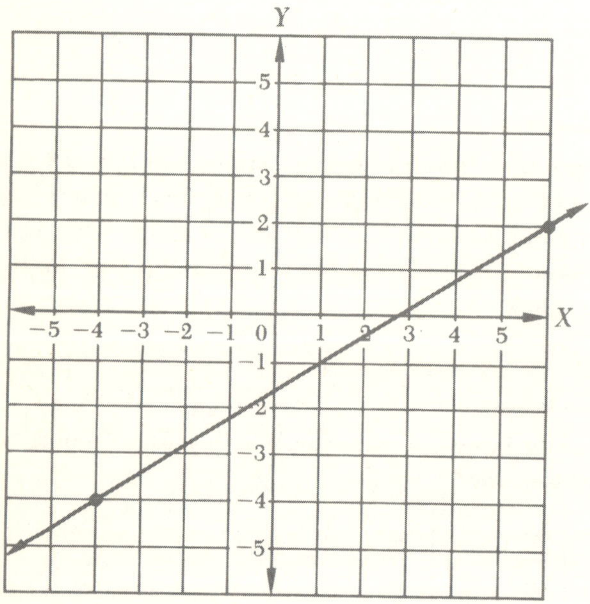

16 *a.* Complete the table of values below for $x + 4y = -3$:

x	0	1	2	3	5
y					

b. Which two points would you choose for graphing the solution set of $x + 4y = -3$?

c. Graph the solution set of $x + 4y = -3$ and check to see that the other points are on the line.

---

a.

x	0	1	2	3	5
y	$-\frac{3}{4}$	-1	$-\frac{5}{4}$	$-\frac{6}{4}$	-2

b. (1,−1); (5,−2)

c.

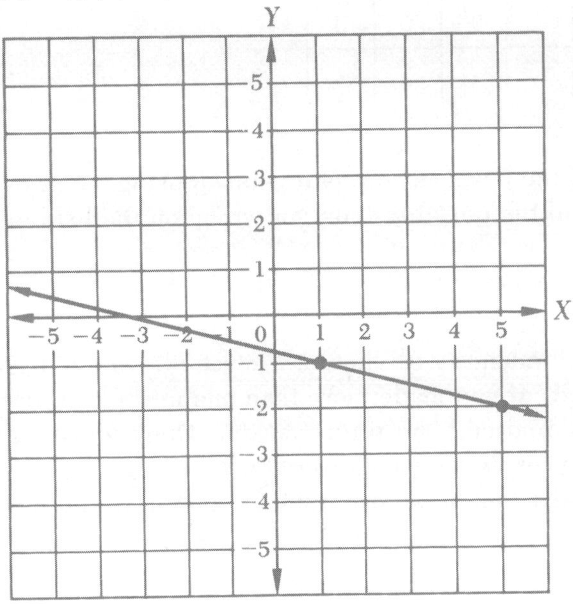

You have finished the program for Sec. 5-3. Now take the posttest to see how much you have learned.

POSTTEST 5-3

For each of the following equations complete the table of values and then graph the solution set:

1. $3x + 4y = 2$

x	-2	-1	0	1	2	3
y						

2. $2x - 5y = -4$

x	-2	-1	0	1	2	3
y						

3. $7x + 3y = -5$

x	-2	-1	0	1	2	3	4
y							

When you have finished the posttest, correct it yourself using the answers on page 277. Then check all the questions you got wrong on the list below:

1. ______ 3. ______

2. ______

If you compare your performance with that on Pretest 5-3, you will see how much you have improved. If you made more than one mistake on this posttest, review the material and take the pretest again. Don't move on to the next section until you can get all the questions right.

Now proceed to the next section whose number you circled on page 196. If no other sections were circled, take Final Test 5 (page 271).

Section 5-4
The Slope-Intercept Form

Do this section if you made a mistake in Questions 7 to 10 of Diagnostic Test 5. Begin by taking the pretest below, which will show you how much of Sec. 5-4 you already know and what you will be able to do by the time you have finished it.

PRETEST 5-4

Without graphing give the slopes and the Y intercepts of the following lines:

1. $3x + 4y = 6$
   Slope = ______
   Y intercept = ______

2. $2x - 5y = -2$
   Slope = ______
   Y intercept = ______

Draw lines through the point $(0, -1)$ with the following slopes:

3. -3
4. $\frac{2}{3}$

Use the slope-intercept method to graph the following equations:

5. $3x = 8 + 4y$
6. $2x + 5y - 7 = 0$

When you have finished the pretest, correct it yourself using the answers on page 278. Then check all the questions you got wrong on the list below:

1. ______	3. ______	5. ______
2. ______	4. ______	6. ______

If you made any mistakes (even if it was only *one* mistake), read the introduction and go through the program.

If you made no mistakes at all, skip Sec. 5-4, because you already know its material, and go on to Sec. 5-5.

Introduction

You will see that first-degree equations in two variables can usually be written in the form $y = mx + c$, where m is the slope and c is the Y intercept, the point of intersection with the Y axis. You learn how the slope of a line can be deduced from its graph and how to graph a line of given slope through a given point. You apply this in graphing equations of the general type $ax + by = c$ by the slope-intercept method.

FRAMES FOR SECTION 5-4

1 First-degree equations in two variables can usually be written in the form $y = mx + c$. Thus

$$3x - y = 6$$

$\updownarrow$ AT with $-3x$

$$-y = 6 - 3x$$

$\updownarrow$ MT with -1

$$y = -6 + 3x$$

In this case $m = 3$, $c = -6$. Similarly

a. $x - 2y = 5$

$\updownarrow$ AT with $-x$

____

$\updownarrow$ MT with $-\frac{1}{2}$

____

giving $m = \dfrac{1}{2}$, $c =$ ______.

b. $2x + 5y = 7$

$\updownarrow$ AT with $-2x$

____

$\updownarrow$ MT with $\frac{1}{5}$

____

giving $m =$ ______, $c =$ ______.

---

a. $-2y = 5 - x$

$$y = -\frac{5}{2} + \frac{x}{2};\ m = \frac{1}{2};\ c = -\frac{5}{2}$$

b. $5y = 7 - 2x$

$$y = \frac{7}{5} - \frac{2x}{5};\ m = -\frac{2}{5};\ c = \frac{7}{5}$$

---

2 Transform the following equations to the form $y = mx + c$ and give the values of m and c:

a. $3x + 4y = 8$

$\updownarrow$ AT with $-3x$

____

$\updownarrow$ MT with $\frac{1}{4}$

____

giving $m =$ ______, $c =$ ______.

b. $4x + 2y + 5 = 0$

$\updownarrow$ AT with $-2y$

____

$\updownarrow$ MT with $-\frac{1}{2}$

____

giving $m =$ ______, $c =$ ______.

c. $3x = 4y - 7$

$\updownarrow$ AT with 7

____

$\updownarrow$ MT with $\frac{1}{4}$

____

giving $m =$ ______, $c =$ ______.

---

a. $4y = 8 - 3x$

$$y = 2 - \frac{3x}{4}$$

$m = -\frac{3}{4}$, $c = 2$

b. $4x + 5 = -2y$

$$-2x - \frac{5}{2} = y$$

$m = -2$; $c = -\frac{5}{2}$

c. $3x + 7 = 4y$

$$\frac{3x}{4} + \frac{7}{4} = y$$

$m = \frac{3}{4}$; $c = \frac{7}{4}$

---

3 Transform the following equations to the form $y = mx + c$ and give the values of m and c:

a. $5x = 3y - 6$

$\updownarrow$ AT with ____

____

$\updownarrow$ MT with ____

____

$m =$ ____, $c =$ ____.

b. $2x - 4y = 7$

$\updownarrow$ AT with ____

____

$\updownarrow$ MT with ____

____

$m =$ ____, $c =$ ____.

c. $5x + 2y + 9 = 0$

$\updownarrow$ AT with ____

____

$\updownarrow$ MT with ____

____

$m =$ ____, $c =$ ____.

---

a.

$\updownarrow$ AT with 6

$5x + 6 = 3y$

$\updownarrow$ MT with $\frac{1}{3}$

$\frac{5x}{3} + 2 = y$

$m = \frac{5}{3}$; $c = 2$

b.

$\updownarrow$ AT with $-2x$

$-4y = 7 - 2x$

$\updownarrow$ MT with $-\frac{1}{4}$

$y = -\frac{7}{4} + \frac{x}{2}$

$m = \frac{1}{2}$, $c = -\frac{7}{4}$

c.

$\updownarrow$ AT with $-2y$

$$5x + 9 = -2y$$

$\updownarrow$ MT with $-\frac{1}{2}$

$$-\frac{5x}{2} - \frac{9}{2} = y$$

$$m = -\frac{5}{2};\ c = -\frac{9}{2}$$

---

4 Transform the following equations to the form $y = mx + c$ and give the values of m and c:

a. $4x = 9 - 3y$
b. $3x + 6y + 4 = 0$
c. $2x - 3y = -7$

---

a.

$$4x = 9 - 3y$$
$$4x - 9 = -3y$$
$$-\frac{4}{3}x + 3 = y$$
$$m = -\frac{4}{3};\ c = 3$$

b.

$$3x + 6y + 4 = 0$$
$$3x + 4 = -6y$$
$$-\frac{x}{2} - \frac{2}{3} = y$$
$$m = -\frac{1}{2};\ c = -\frac{2}{3}$$

c.

$$2x - 3y = -7$$
$$-3y = -7 - 2x$$
$$y = \frac{7}{3} + \frac{2x}{3}$$
$$m = \frac{2}{3};\ c = \frac{7}{3}$$

---

5 *a.* Complete the table of values for $x - 2y = 5$:

x	1	2	3	4	5	6
y						

b. Complete the table of values for $2x - 4y = 7$:

x	1	2	3	4	5	6
y						

c. Graph the solution sets of $x - 2y = 5$ and $2x - 4y = 7$ on the same coordinate system.

---

a.

x	1	2	3	4	5	6
y	-2	$-\frac{3}{2}$	-1	$-\frac{1}{2}$	0	$\frac{1}{2}$

b.

x	1	2	3	4	5	6
y	$-\frac{5}{4}$	$-\frac{3}{4}$	$-\frac{1}{4}$	$+\frac{1}{4}$	$\frac{3}{4}$	$\frac{5}{4}$

c.

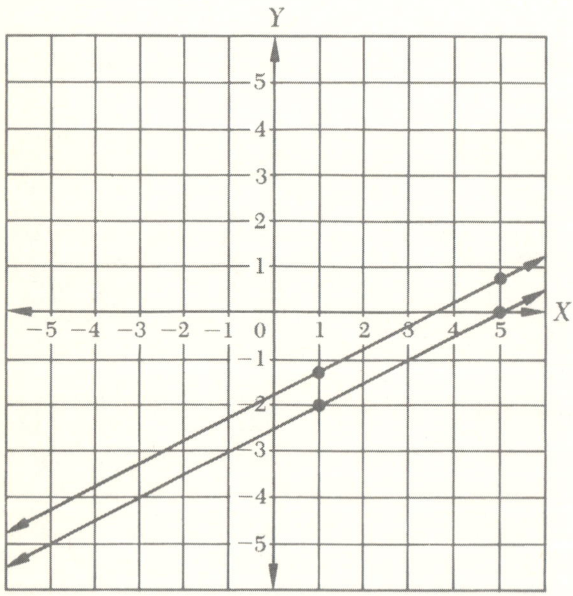

You need not have used the same points, but check carefully that you graphed the right lines.

6 Consider the bottom line $x - 2y = 5$.

a. If you start at (1,−2) and go to (3,−1), you move ______ square(s) upward and ______ square(s) to the right.
b. If you start at (3,−1) and go to (5,0), you move ______ squares(s) upward and ______ square(s) to the right.
c. If you start anywhere on the line and move 2 squares to right along the line, you move ______ square(s) upward.

---

a. 1, 2 *b.* 1; 2 *c.* 1

---

7 *a.* If you start anywhere on the line $x - 2y = 5$ and move 1 square to the right along the line, you move ______ square(s) upward.
b. What is the m value for $x - 2y = 5$ (see Frame 1*b*)?
c. What is the m value for $2x - 4y = 7$ (see Frame 3*b*)?
d. Did you notice anything special about the two lines you drew in Frame 5?

---

a. $\frac{1}{2}$ *b.* $\frac{1}{2}$ *c.* $\frac{1}{2}$ *d.* They are parallel.

---

8 *a.* Complete the table of values for $3x - y = 6$:

x	0	1	2	3	4	5
y						

b. Graph the solution set of $3x - y = 6$.

---

a.

x	0	1	2	3	4	5
y	−6	−3	0	3	6	9

b.

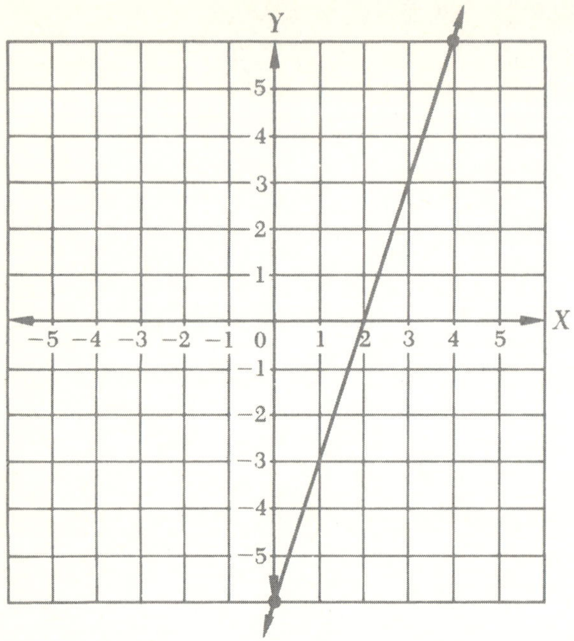

You need not have used the same points, but check carefully that you graphed the right line.

---

9 *a.* If you start anywhere on the line $3x - y = 6$ and move 1 square to the right along the line, you move ______ squares upward.

b. What is the m value for $3x - y = 6$ (see Frame 1*a*)?

---

a. 3 *b.* 3

---

10 *a.* Complete the table of values for $4x = 9 - 3y$:

x	0	1	2	3	4	5	6
y							

b. Graph the solution set of $4x = 9 - 3y$.

---

a.

x	0	1	2	3	4	5	6
y	3	$\frac{5}{3}$	$\frac{1}{3}$	-1	$-\frac{7}{3}$	$-\frac{11}{3}$	-5

b.

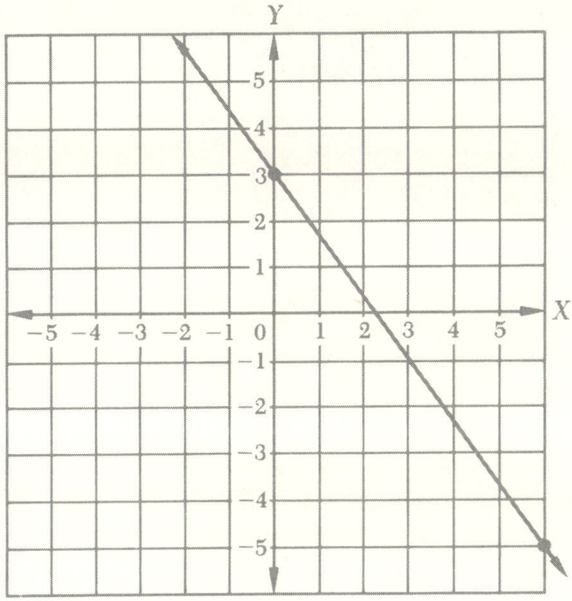

You need not have used the same points, but check carefully that you graphed the right line.

---

11 *a.* If you start anywhere on the line $4x = 9 - 3y$ and move 1 square to the right along the line, you move ______ squares downward, i.e., −______ squares upward.

b. What is the m value for $4x = 9 - 3y$ (see Frame 4*a*)?

---

a. $\frac{4}{3}$; $-\frac{4}{3}$ *b.* $-\frac{4}{3}$

---

12 Complete the table of values for $3x + 4y = 8$:

x	0	1	2	3	4	5
y						

and graph the solution set of $3x + 4y = 8$.

---

x	0	1	2	3	4	5
y	2	$\frac{5}{4}$	$\frac{1}{2}$	$-\frac{1}{4}$	-1	$-\frac{7}{4}$

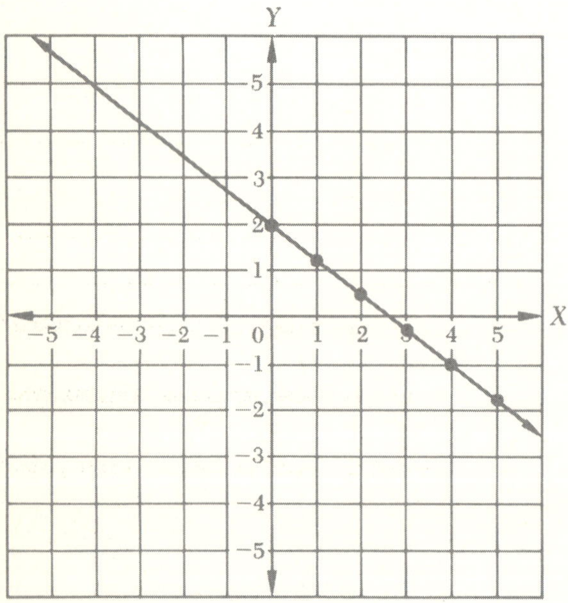

---

13 *a.* If you start anywhere on the line $3x + 4y = 8$ and move 1 square to the right along the line, you move ______ square downward, i.e., ______ square upward.

b. What is the m value for $3x + 4y = 8$ (see Frame 2*a*)?

---

a. $\frac{3}{4}$; $-\frac{3}{4}$ *b.* $-\frac{3}{4}$

---

14 In Frames 7, 9, 11, and 13 you may have noticed that if m is positive, you move (upward/downward) __________ as you move to the right along the line and if m is negative, you move (upward/downward) __________ as you move to the right along the line.

upward; downward

15 For which of the lines below is m positive?

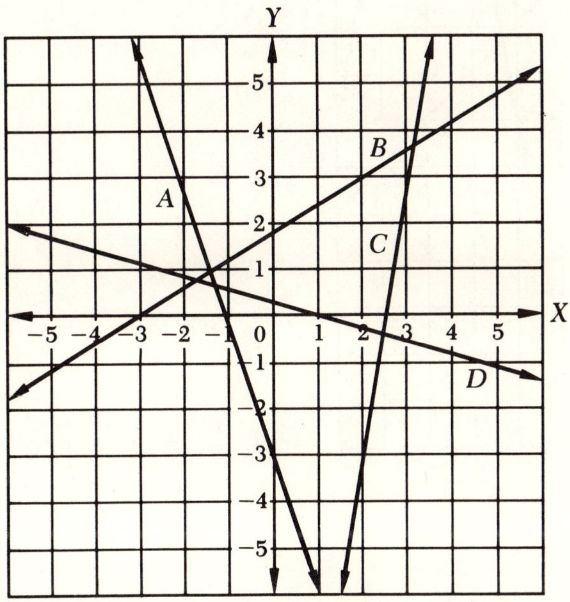

B and *C*

16 The m in $y = mx + c$ is called the *slope* or *gradient* of the line. It represents the number of squares you move upward or (if negative) downward for every square you move to the right along the line. Give the slopes of lines A, B, C and D below:

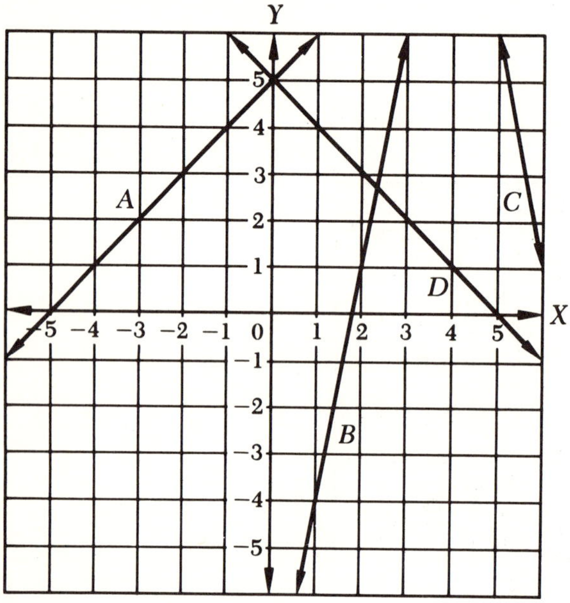

$A = 1$ $B = 5$ $C = -5$ $D = -1$

---

17 Give the slopes of lines A, B, and C in the diagram below:

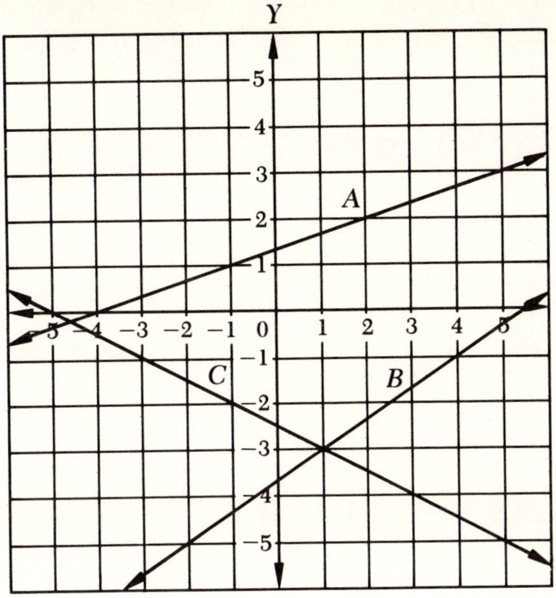

$A = \frac{1}{3}$ $B = \frac{2}{3}$ $C = -\frac{1}{2}$

18 Draw lines going through the origin which have the following gradients:
$A = 4$ $B = \frac{1}{2}$ $C = -\frac{1}{3}$ $D = -3$

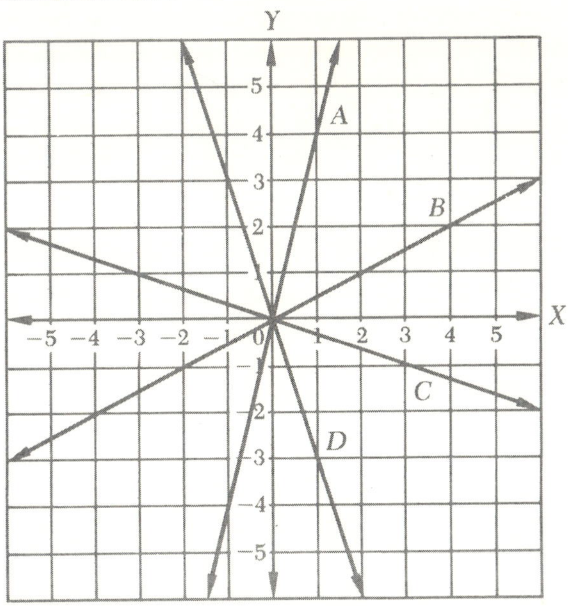

19 Draw lines going through the origin which have the following gradients:
$A = 3 \quad B = \frac{1}{4} \quad C = -\frac{1}{4} \quad D = -1$

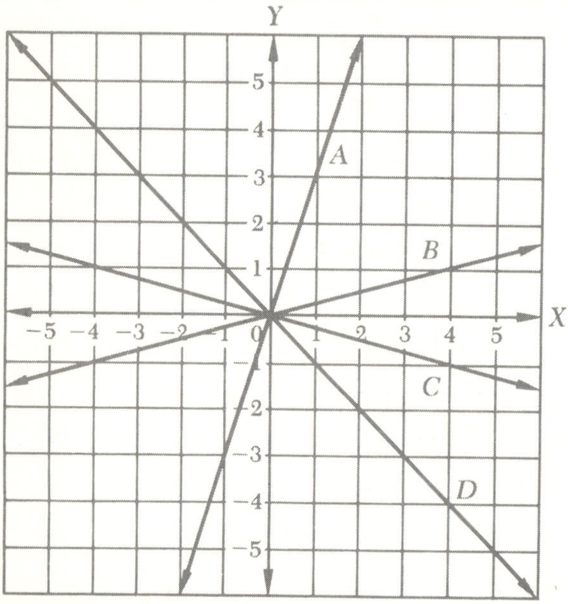

20 Draw lines going through the origin which have the following gradients:
$A = \frac{3}{2} \quad B = \frac{3}{4} \quad C = -\frac{2}{3} \quad D = -\frac{5}{3}$

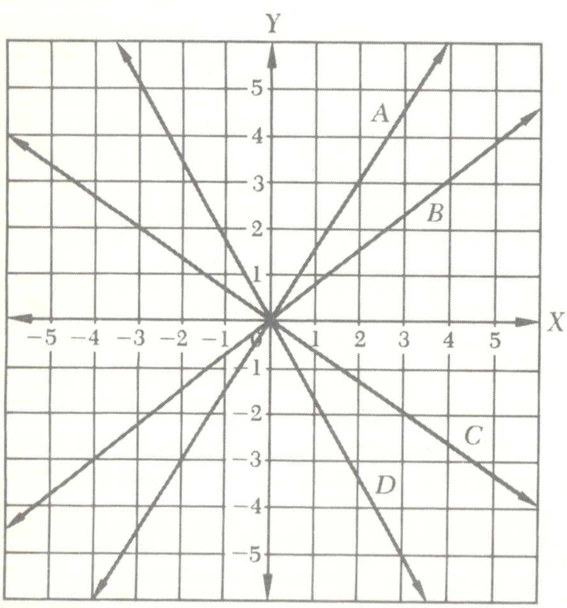

21 Draw a line through the point $(0,-2)$ with slope $\frac{3}{2}$.

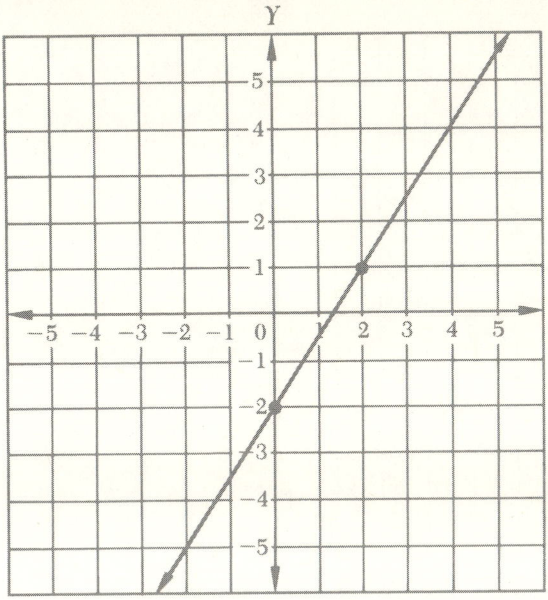

Check very carefully that you have drawn your line correctly.

22 Draw a line through the point $(0,2)$ with slope $-\frac{2}{3}$.

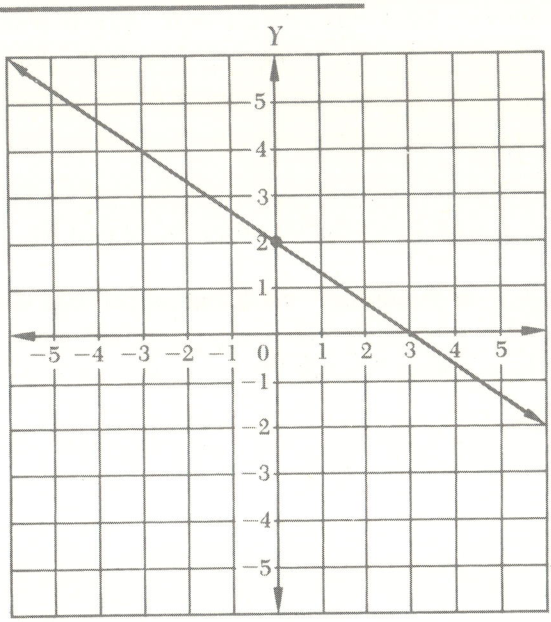

Check very carefully that you have drawn your line correctly.

23 Draw two lines through the point $(0,-3)$ with slopes of $\frac{5}{2}$ and $-\frac{3}{5}$.

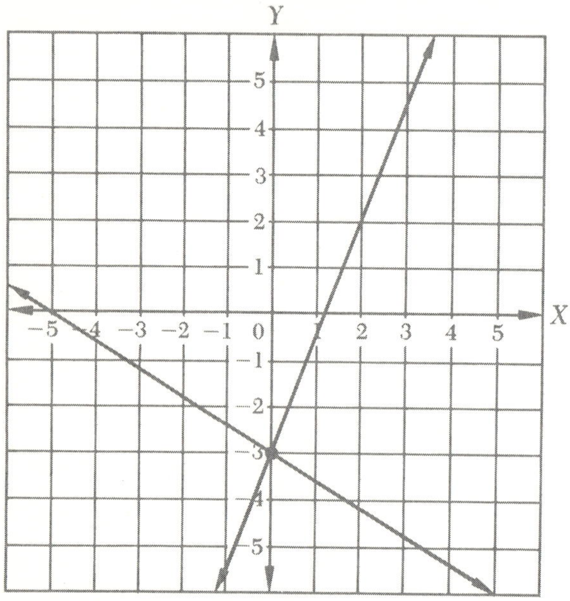

24 We have already seen that any first-degree equation in two variables can be written in the form $y = mx + c$ and that m is the __________ of the graph. The graph will intersect the Y axis when $x = 0$. At this point $y = m(0) + c =$ ______. Thus the graph of $y = mx + c$ intersects the Y axis at the point (__,__).

slope (or gradient); c; $(0,c)$

25 *a.* The graph of $y = mx + c$ goes through the point (0,__) and has a slope of ______.

b. The graph of $y = 3x + 1$ goes through the point (0,__) and has a slope of ______.

c. And the graph of $y = 2x - 3$ goes through the point (0,__) and has a slope of ______.

a. c; m *b.* 1; 3 *c.* −3; 2

26 Use this information to draw the graphs of $y = 3x + 1$ and $y = 2x - 3$ on the same coordinate system.

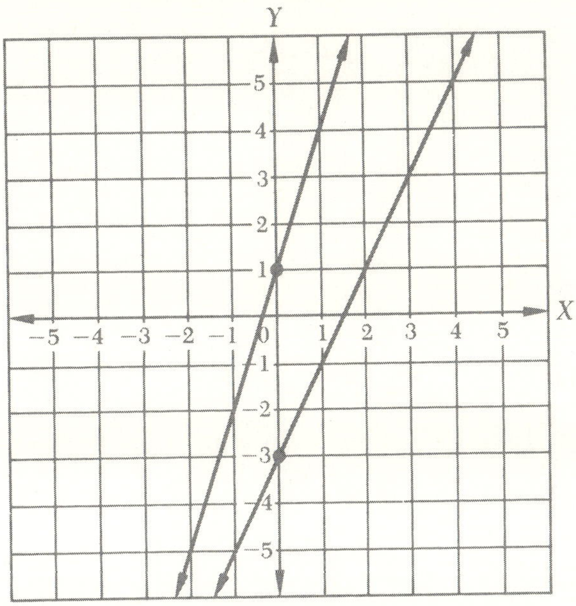

---

27 The point at which the graph intersects the Y axis is called the Y intercept. What are the Y intercepts and the slopes of the following graphs?

a. $y = 3x + 1$
b. $y = 2x - 3$
c. $y = -5x + 4$
d. $y = mx + c$

---

a. (0, 1); 3 *b.* (0, −3); 2 *c.* (0, 4); −5 *d.* $(0, c)$; m

---

28 In Frame 26 you graphed two equations by the *slope-intercept method.* (First you found the Y intercept, then drew a line through that point with the correct slope.) Use the slope-intercept method to graph

$$y = \frac{2x}{3} + 2$$

(slope = ______; Y intercept = ______).

$\frac{2}{3}$ $\quad$ (0,2)

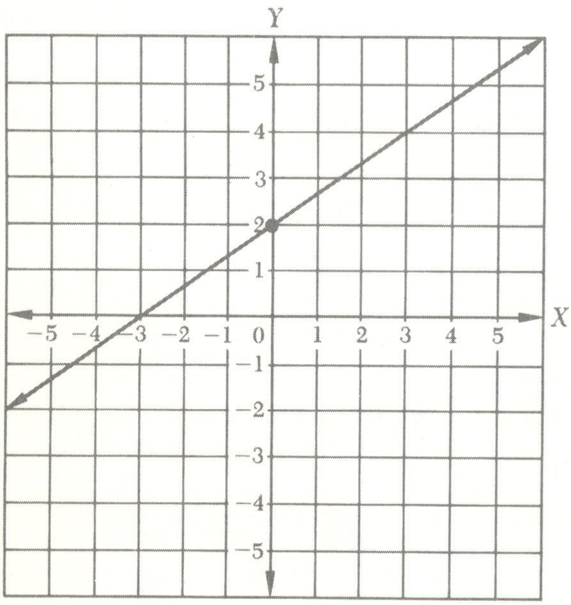

---

29 Use the slope-intercept method to graph

$$y = -\frac{1}{4}x + \frac{1}{2} \qquad \text{and} \qquad y = \frac{5x}{2} - \frac{3}{2}$$

on the same coordinate system.

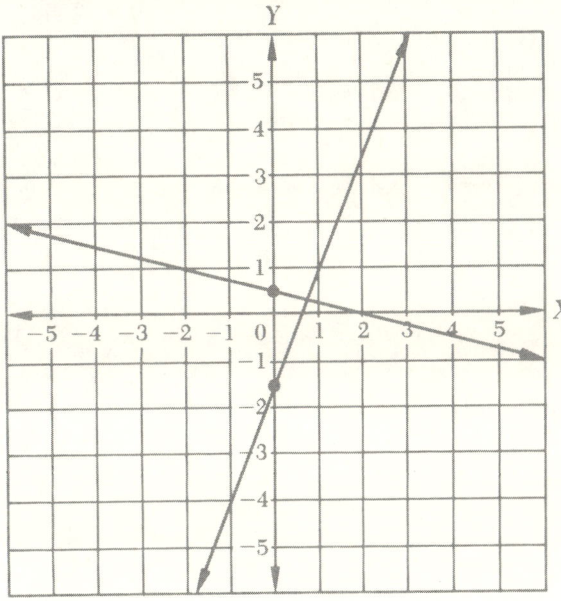

30 Convert the equation $2x = 3 - 4y$ to the form $y = mx + c$ and graph it by the slope-intercept method. ($y =$ ________; slope = ______; Y intercept = ______.)

$-\frac{1}{2}x + \frac{3}{4}$ $-\frac{1}{2}$ $\left(0, \frac{3}{4}\right)$

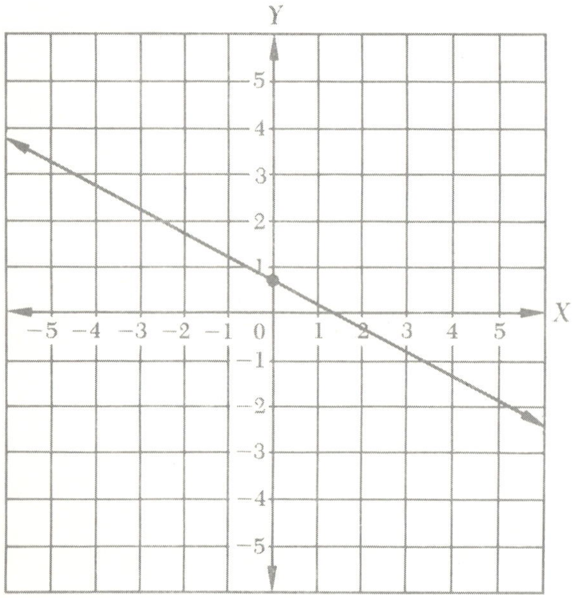

---

31 Convert the following equations to the form $y = mx + c$ and graph them by the slope-intercept method:

a. $3x = 5 - y$
b. $3x + 2y + 4 = 0$
c. $x - 2y = -5$

---

a. $y = -3x + 5$

b. $y = -\frac{3x}{2} - 2$

c. $y = \frac{x}{2} + \frac{5}{2}$

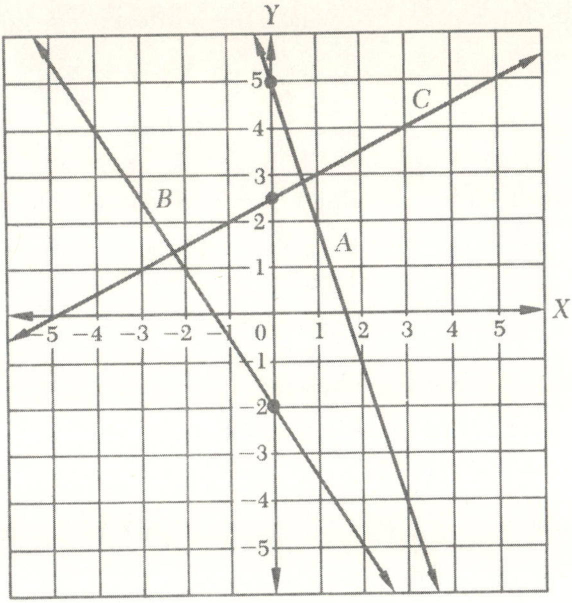

---

32 Graph the following equations by the slope-intercept method:

a $2x = 9 - 3y$
b. $x - 2y + 6 = 0$
c. $5x + 2y = -4$

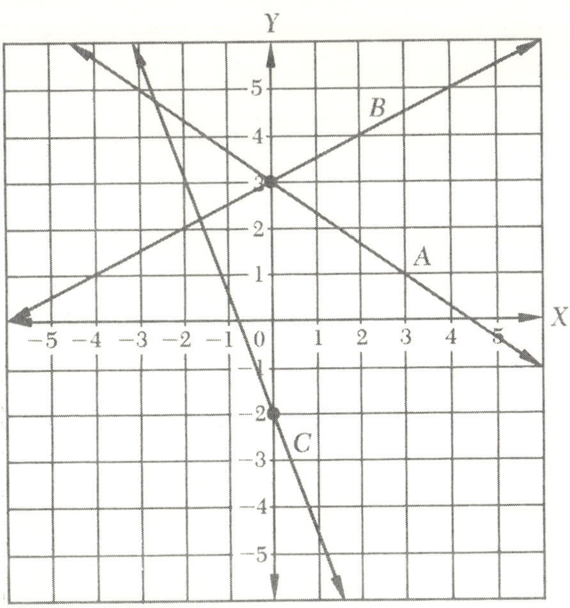

You have finished the program for Sec. 5-4. Now take the posttest to see how much you have learned.

POSTTEST 5-4

Without graphing give the slopes and the Y intercepts of the following lines:

1. $2x + 5y = 6$
   Slope = ______.
   Y intercept = ______.

2. $3x - 4y = -2$
   Slope = ______.
   Y intercept = ______.

Draw lines through the point $(0,-2)$ with the following slopes:

3. $\frac{3}{2}$

4. $-\frac{1}{3}$

Use the slope-intercept method to graph the following equations:

5. $2x = 8 + 5y$

6. $3x + 4y - 7 = 0$

When you have finished the posttest correct it yourself using the answers on page 278. Then check all the questions you got wrong on the list below:

1. ______	3. ______	5. ______
2. ______	4. ______	6. ______

If you compare your performance with that on Pretest 5-4, you will see how much you have improved. If you made more than one mistake on this posttest, review the material and take the pretest again. Don't move on to the next section until you can get all the questions right.

Now proceed to Sec. 5-5.

Section 5-5
Building Tables of Values from $y(0)$ and $y(1)$ and Graphing $y = k$ and $x = q$

Do this section if you made a mistake in Questions 7 to 10 of Diagnostic Test 5. Begin by taking the pretest below, which will show you how much of Sec. 5-5 you already know and what you will be able to do by the time you have finished it.

PRETEST 5-5

Give the value of m(the slope), $y(0)$, and $y(1)$ for the following lines:

1. $6x + 3y = -7$

   $m =$ ______

   $y(0) =$ ______

   $y(1) =$ ______

2. $4x - 5y = -8$

   $m =$ ______

   $y(0) =$ ______

   $y(1) =$ ______

Complete the following tables of values and graph the equations from which they are derived, assuming that they are straight lines:

3.

x	0	1	2	3	4	5
y	$\frac{1}{2}$	$\frac{5}{6}$				

4.

x	0	1	2	3	4	5
y	$-\frac{7}{3}$	$-\frac{11}{3}$				

Graph the following equations:

5. $y = -3$

6. $x = 4$

When you have finished the pretest, correct it yourself using the answers on page 279. Then check all the questions you got wrong on the list below:

1. ______ 3. ______ 5. ______

2. ______ 4. ______ 6. ______

If you made any mistakes (even if it was only *one* mistake), read the introduction and go through the program.

If you made no mistakes at all, skip Sec. 5-5, because you already know its material, and go on to take Final Test 5 (page 271).

Introduction

After learning that the notation $y(n)$ represents the value of y at $x = n$, you are shown that $y(n) = y(n-1) + m$. By finding $y(0)$ and $y(1)$ by substitution, you can easily calculate m and hence $y(2)$, $y(3)$, $y(4)$, ... You are then shown how two integral values of $y(n)$ can be used as a basis for checking and for graphing an accurate straight line. This method is simple and accurate, as $y(0)$ and $y(1)$ are usually easy to find. Some special cases of lines parallel to the X or Y axis, such as $x = w$ and $y = k$, are also discussed.

FRAMES FOR SECTION 5-5

1 Most first-degree equations in two variables can be written in the form $y = mx + c$, where m is the ______ of their graph.

a. At $x = 0, y = m(0) + c =$ ______.
b. At $x = 1, y = m(1) + c =$ ______.
c. At $x = 2, y =$ ______.
d. At $x = 3, y =$ ______.

---

slope (or gradient) *a.* c *b.* $m + c$ *c.* $2m + c$ *d.* $3m + c$

---

2 If we say that the value of y at $x = n$ is called $y(n)$ the value of y at $x = 0$ is $y(0) =$ ______; the value of y at $x = 1$ is $y(1) =$ ______; and $y(2) =$ ______.

---

c $m + c$ $2m + c$

---

3 Similarly:

a. $y(3) =$ ______
b. $y(4) =$ ______
c. $y(n) =$ ______

---

a. $3m + c$ *b.* $4m + c$ *c.* $nm + c$

---

4 $y(1) - y(0) =$ ______ $-$ ______ $=$ ______; therefore $y(1) = y(0) +$ ______. Similarly $y(2) = y(1) +$ ______, $y(3) = y(2) +$ ______, $y(4) = y(3) +$ ______, and so forth.

$(m + c) - c = m$ m m m m

---

5 We can find $y(0)$ and $y(1)$ by substituting 0 and 1 for x in the equation concerned. Thus for the equation $5x - 3y = 7$:

a. $5(0) - 3y = 7 \longleftrightarrow y = -\frac{7}{3}$, and so $y(0) =$ ______.

b. $5(1) - 3y = 7 \longleftrightarrow y =$ ______, and so $y(1) =$ ______.

c. Hence $m = y(1) - y(0) =$ ______.

---

a. $-\frac{7}{3}$ *b.* $-\frac{2}{3}; -\frac{2}{3}$ *c.* $\frac{5}{3}$

---

6 *a.* $y(2) = y(1) + m = -\frac{2}{3} + \frac{5}{3} =$ ______

b. $y(3) = y(2) + m =$ ______ $+ \frac{5}{3} =$ ______

c. Find $y(4)$ and $y(5)$ in a similar way and complete the table of values:

x	0	1	2	3	4	5
y	$-\frac{7}{3}$	$-\frac{2}{3}$				

d. At which of these points are x and y both integral?

---

a. $\frac{3}{3}$ or 1 *b.* $\frac{3}{3} + \frac{5}{3} = \frac{8}{3}$

c.

x	0	1	2	3	4	5
y	$-\frac{7}{3}$	$-\frac{2}{3}$	$\frac{3}{3} = 1$	$\frac{8}{3}$	$\frac{13}{3}$	6

d. (2, 1) and (5, 6)

---

7 Check that (2, 1) and (5, 6) are solutions of $5x - 3y = 7$. Then use these two points to graph $5x - 3y = 7$.

CHECK

$5(2) - 3(1) = 10 - 3 = 7$
$5(5) - 3(6) = 25 - 18 = 7$

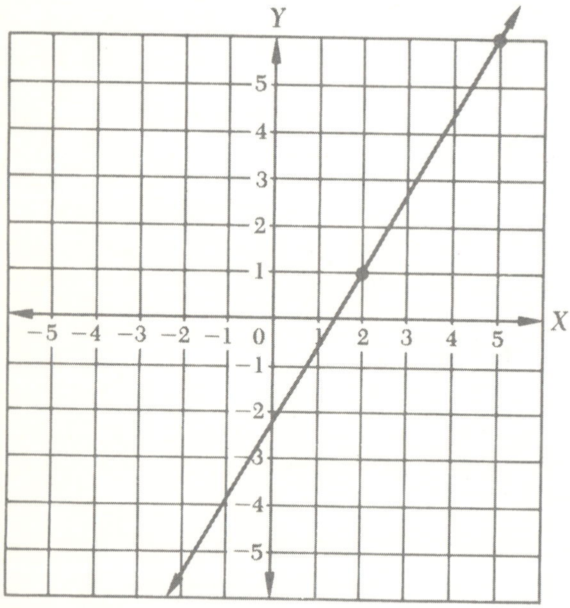

---

8 Now let us consider the equation $3x + 4y = 7$:

a. $y(0) =$ ______ because $3(0) + 4y = 7 \longleftrightarrow y =$ ______.
b. $y(1) =$ ______ because $3(1) + 4y = 7 \longleftrightarrow y =$ ______.
c. Therefore $m = y(1) - y(0) =$ ______.
d. $y(2) = y(1)$; $m =$ ______.
e. $y(3) = y(2)$; $m =$ ______.

---

a. $\frac{7}{4}; \frac{7}{4}$ *b.* 1; 1 *c.* $-\frac{3}{4}$ *d.* $\frac{1}{4}$ *e.* $-\frac{1}{2}$ or $-\frac{2}{4}$

---

9 *a.* Complete the table of values for $3x + 4y = 7$:

x	0	1	2	3	4	5
y						

b. Check two points at which x and y are both integers.

c. Graph the equation $3x + 4y = 7$.

---

a.

x	0	1	2	3	4	5
y	$\frac{7}{4}$	$\frac{4}{4} = 1$	$\frac{1}{4}$	$-\frac{2}{4}$	$-\frac{5}{4}$	$-\frac{8}{4} = -2$

b. For (1,1): $3(1) + 4(1) = 3 + 4 = 7$

For (5,−2): $3(5) + 4(-2) = 15 + (-8) = 7$

c.

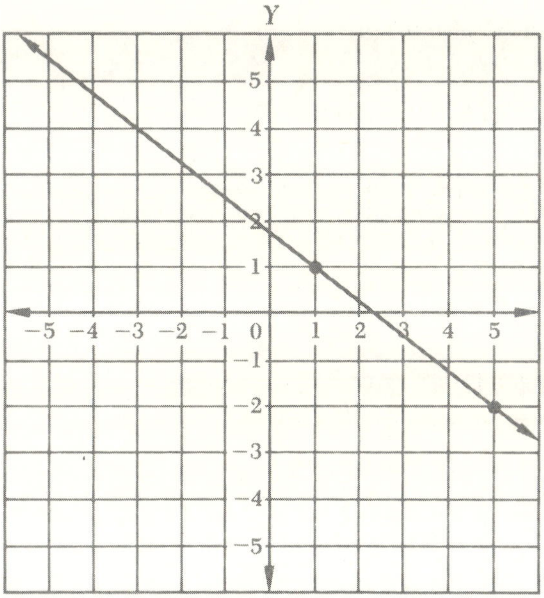

---

10 We have used two methods of graphing a first-degree equation in two variables, the *slope-intercept* method and the *table-of-values* method. A third method combines the two by using the slope to construct a table of values. Thus, with $2x - 3y = 8$:

a. $y(0) =$ ______.
$y(1) =$ ______.
Slope, $m = y(1) - y(0) =$ ______.

b. $y(2) = y(1) + m =$ ______.
$y(3) = y(2) + m =$ ______, etc.

c. Complete the table of values in this way:

x	0	1	2	3	4	5
y						

d. Check two points and graph the line.

---

a. $-\frac{8}{3}$; -2; $m = +\frac{2}{3}$ *b.* $y(2) = -\frac{4}{3}$; $y(3) = -\frac{2}{3}$

c.

x	0	1	2	3	4	5
y	$-\frac{8}{3}$	$-\frac{6}{3} = -2$	$-\frac{4}{3}$	$-\frac{2}{3}$	0	$\frac{2}{3}$

d. Checking $(1,-2)$: $2(1) - 3(-2) = 2 + 6 = 8$;
checking $(4,0)$: $2(4) - 3(0) = 8$

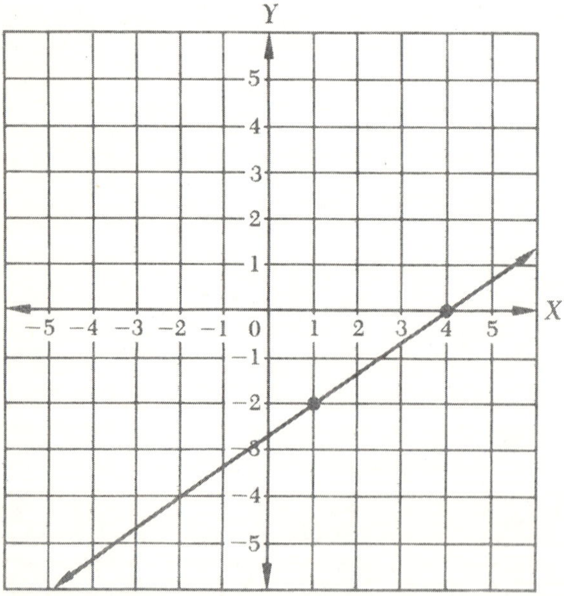

---

11 Use the same method to graph $4x + 3y = -3$:

a. $y(0) =$ ______.
$y(1) =$ ______.
$m =$ ______.

b. $y(2) =$ ______.
$y(3) =$ ______, etc.

c. Complete the table of values in this way:

x	0	1	2	3	4	5
y						

d. Check two points and graph the line.

---

a. $y(0) = -1$; $y(1) = -\frac{7}{3}$; $m = -\frac{4}{3}$

b. $y(2) = -\frac{11}{3}$; $y(3) = -\frac{15}{3} = -5$

c.

x	0	1	2	3	4	5
y	$-\frac{3}{3} = -1$	$-\frac{7}{3}$	$-\frac{11}{3}$	$-\frac{15}{3} = -5$	$-\frac{19}{3}$	$-\frac{23}{3}$

d. Checking $(0, -1)$: $4(0) + 3(-1) = -3$;
checking $(3, -5)$: $4(3) + 3(-5) = 12 + (-15) = -3$

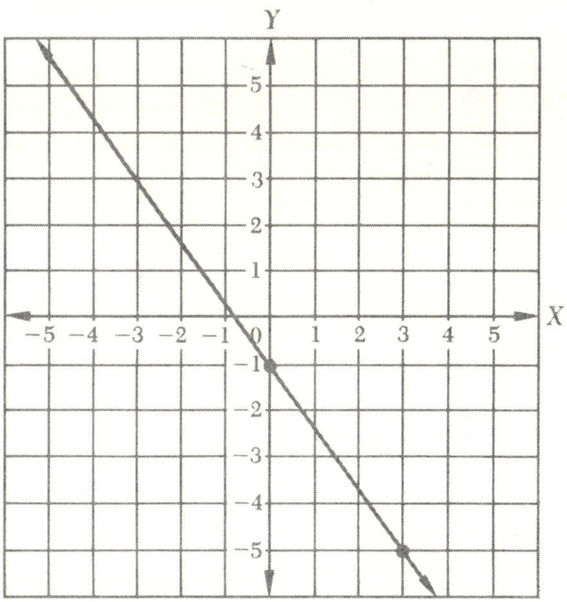

12 Find the slope of the following line and complete the table of values:

x	0	1	2	3	4	5
y	3	$\frac{5}{2}$				

---

slope $= -\frac{1}{2}$

x	0	1	2	3	4	5
y	3	$\frac{5}{2}$	2	$\frac{3}{2}$	1	$\frac{1}{2}$

---

13 Find the slopes of the following lines and complete the tables of values:

a.

x	0	1	2	3	4	5
y	$-\frac{11}{5}$	$-\frac{7}{5}$				

b.

x	0	1	2	3	4	5
y	$\frac{11}{4}$	2				

---

a. slope $= \frac{4}{5}$

x	0	1	2	3	4	5
y	$-\frac{11}{5}$	$-\frac{7}{5}$	$-\frac{3}{5}$	$+\frac{1}{5}$	$\frac{5}{5} = 1$	$\frac{9}{5}$

b. slope $= -\frac{3}{4}$

x	0	1	2	3	4	5
y	$\frac{11}{4}$	$\frac{8}{4} = 2$	$\frac{5}{4}$	$\frac{2}{4} = \frac{1}{2}$	$-\frac{1}{4}$	$-\frac{4}{4} = -1$

---

14 *a.* Use the method of the previous frames to graph $5x + 2y - 3 = 0$. $y(0) =$ ______; $y(1) =$ ______; $m =$ ______.

b.

x	0	1	2	3	4	5
y						

c. Then check two points and graph the line.

a. $+\frac{3}{2}$; -1; $-\frac{5}{2}$

b.

x	0	1	2	3	4	5
y	$+\frac{3}{2}$	$-\frac{2}{2} = -1$	$-\frac{7}{2}$	$-\frac{12}{2} = -6$	$-\frac{17}{2}$	$-\frac{22}{2} = -11$

c. Checking $(1,-1)$: $5(1) + 2(-1) - 3 = 5 + (-2) - 3 = 0$;
checking $(3,-6)$: $5(3) + 2(-6) - 3 = 15 + (-12) - 3 = 0$

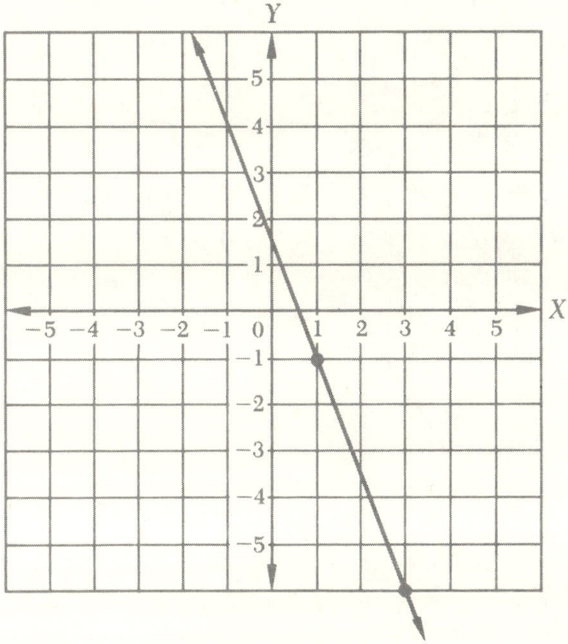

15 *a.* Use the same method to graph $3x - 2y - 4 = 0$. $y(0) =$ ______; $y(1) =$ ______; $m =$ ______.

b.

x	0	1	2	3	4	5
y						

c. Then check two points and graph the line.

---

a. $-2; -\frac{1}{2}; \frac{3}{2}$

b.

x	0	1	2	3	4	5
y	-2	$-\frac{1}{2}$	1	$\frac{5}{2}$	4	$\frac{11}{2}$

c. Checking $(0,-2)$: $3(0) - 2(-2) - 4 = 4 - 4 = 0$;
checking $(4,4)$: $3(4) - 2(4) - 4 = 12 - 8 - 4 = 0$

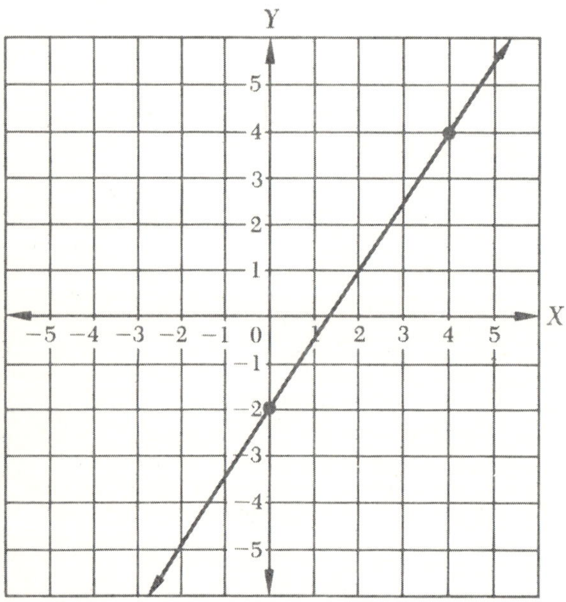

---

16 Use the same method to graph $3x - y - 2 = 0$. (Refer back to Frame 15 if you cannot remember the method.)

$y(0) = -2$; $y(1) = 1$; $m = 3$;

x	0	1	2	3	4	5
y	−2	1	4	7	10	13

Checking $(0,-2)$: $3(0) - (-2) - 2 = 2 - 2 = 0$;
checking $(2,4)$: $3(2) - 4 - 2 = 6 - 4 - 2 = 0$

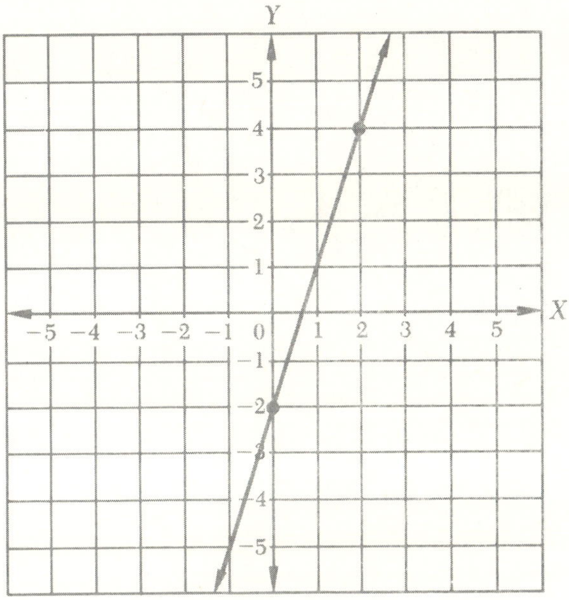

---

17 Use the same method to graph $x - 3y = 4$. (Remember to check two points before graphing.)

$y(0) = -\frac{4}{3}$; $y(1) = -1$; $m = \frac{1}{3}$

x	0	1	2	3	4	5
y	$-\frac{4}{3}$	-1	$-\frac{2}{3}$	$-\frac{1}{3}$	0	$\frac{1}{3}$

Checking (4,0): $4 - 3(0) = 4$; checking (1,−1): $1 - 3(-1) = 1 + 3 = 4$

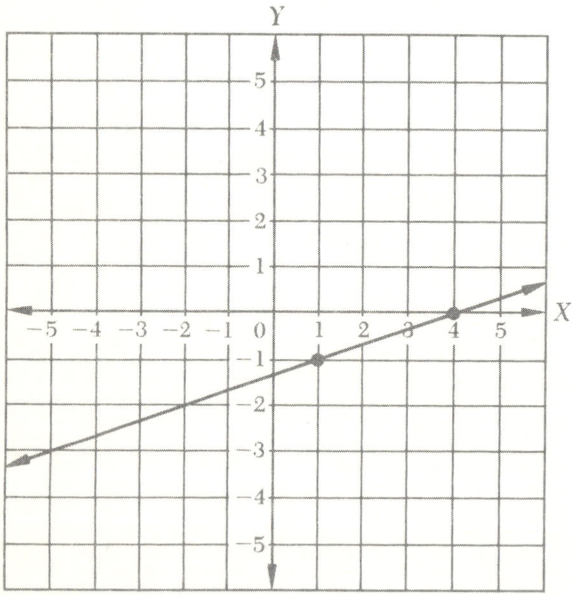

---

18 *a.* If $m = 0$, then $y = mx + c$ becomes $y =$ _____. Thus an equation of the form $y = 4$ which contains no term in x represents a line with a slope of _____.

b. Is (5,4) a solution of $y = 4$?

c. Is $(-5,4)$ a solution of $y = 4$?

d. Graph the equation $y = 4$ in the coordinate plane.

a. $y = c$; 0 *b.* yes *c.* yes

d.

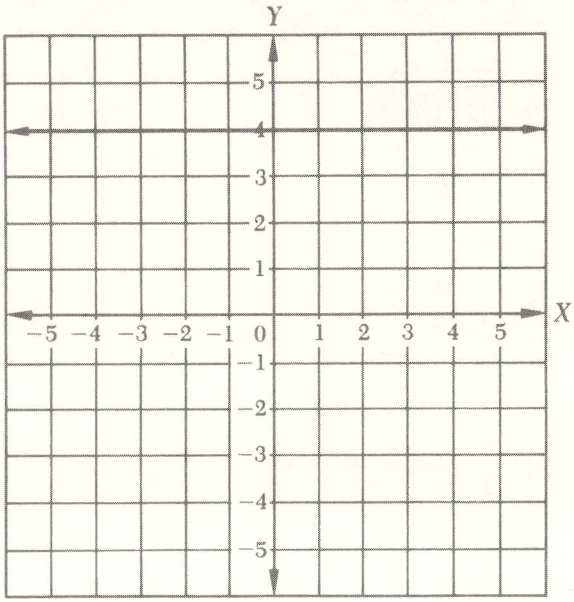

19 Graph the following equations:

A: $y = 2$ B: $y = -1$ C: $y = -4$

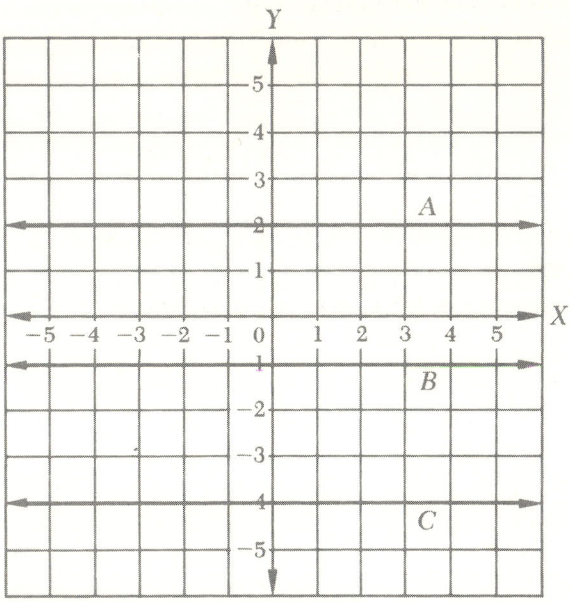

20 Graph the following equations:

A: $y = -2$ B: $x = -2$

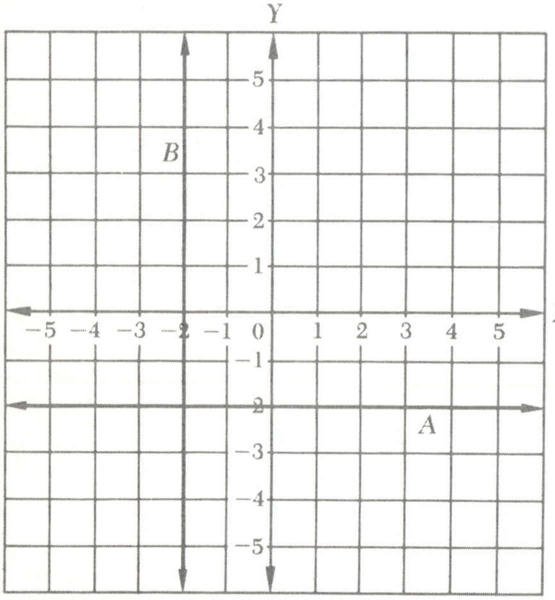

---

21 Since the graph of $x = -2$ does not move to the right at all, and $y(0)$ and $y(1)$ do not exist, it is impossible to define its slope. Give the slopes of the following equations (if the slope is not defined say so):

a. $y = -5$
b. $y - x = -5$
c. $y - 1{,}000x = -5$
d. $-\ 1{,}000x = -5$

---

a. 0 *b.* 1 *c.* 1,000 *d.* not defined

---

22 Give the slopes of the following graphs and then draw them:

A: $x = 5$ B: $x = -3$ C: $y = 1$

A, not defined; B, not defined; C, 0

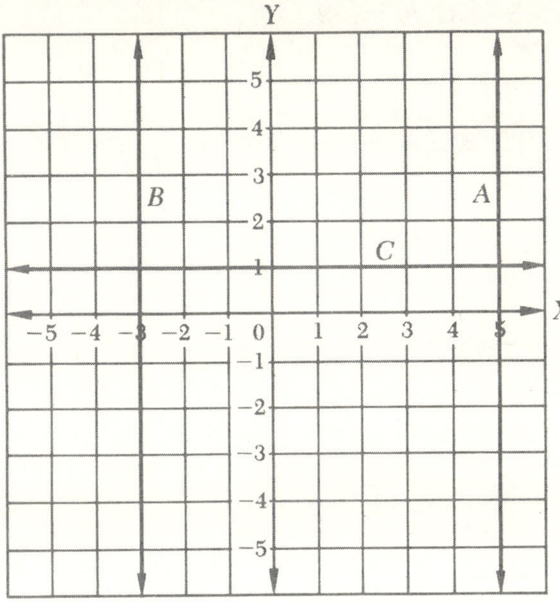

23 *a.* Which of the lines in Frame 22 were parallel to the X axis?
b. What is the equation of the X axis?

a. C ($y = 1$) *b.* $y = 0$

24 *a.* Which of the lines in Frame 22 were parallel to the Y axis?
b. What is the equation of the Y axis?

a. A ($x = 5$) and B ($x = -3$) *b.* $x = 0$

25 Give the slopes of the following lines:

a. $x = -5$
b. $y = 0$
c. $x = 0$
d. $y = 8$

a. not defined *b.* 0 *c.* not defined *d.* 0

26 Graph the following equations by the methods used in this section:

A: $2x + 3y = 4$ B: $x = 3$ C: $y = -2$

A, $y(0) = \frac{4}{3}$; $y(1) = \frac{2}{3}$; $m = -\frac{2}{3}$

x	0	1	2	3	4	5
y	$\frac{4}{3}$	$\frac{2}{3}$	0	$-\frac{2}{3}$	$-\frac{4}{3}$	-2

Checking $(5,-2)$: $2(5) + 3(-2) = 10 + (-6) = 4$;
checking $(2,0)$: $2(2) + 3(0) = 4$

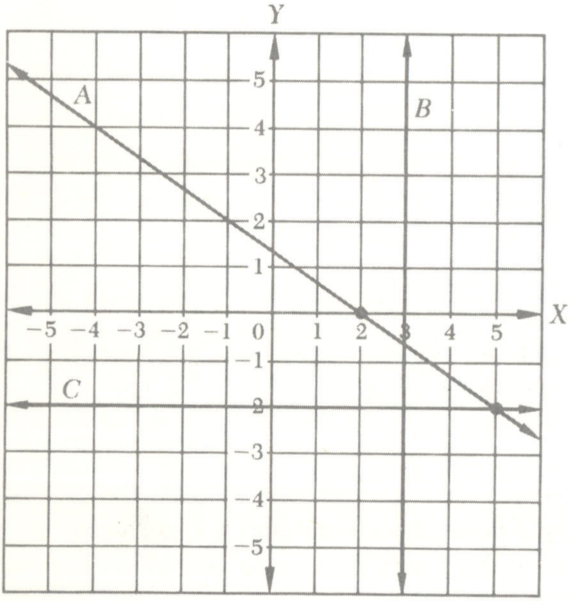

You have finished the program for Sec. 5-5. Now take the posttest to see how much you have learned.

POSTTEST 5-5

Give the values of m (the slope), $y(0)$, and $y(1)$ for the following lines:

1. $3x - 5y = -3$

 $m =$ ______
 $y(0) =$ ______
 $y(1) =$ ______

2. $5x + 10y = -4$

 $m =$ ______
 $y(0) =$ ______
 $y(1) =$ ______

Complete the following tables of values and graph the equations from which they are derived, assuming that they are straight lines:

3.

x	0	1	2	3	4	5
y	$\frac{7}{6}$	$\frac{1}{2}$				

4.

x	0	1	2	3	4	5
y	$-\frac{5}{3}$	$-\frac{1}{3}$				

Graph the following equations:

5. $y = 5$

6. $x = -4$

When you have finished the posttest correct it yourself using the answers on page 279. Then check all the questions you got wrong on the list below:

1. _____ 3. _____ 5. _____

2. _____ 4. _____ 6. _____

If you compare your performance with that on Pretest 5-5, you will see how much you have improved. If you made more than one mistake on this posttest, review the material and take the pretest again. Don't move on to the next section until you can get all the questions right.

Now proceed to Final Test 5.

FINAL TEST 5

Complete the following statements:

1. $(a,b) = (\_\_,-\frac{5}{4})$ is a solution of $3a + 4b = -5$.
2. $(c,d) = (-7,\_\_)$ is a solution of $5c - 2d = -2$.

Draw a rectangular coordinate system and graph the following ordered pairs:

3. $(\frac{7}{3},-2)$

4. $(-\frac{7}{4},-\frac{5}{2})$

5. Complete the table of values and graph the equation $2x + 3y = -7$:

x	0	1	2	3	4	5
y						

6. Complete the table of values and graph the equation $4x - 3y = 10$:

x	0	1	2	3	4	5
y						

7. Graph $5x - 3y = -4$ by the slope-intercept method.

8. Complete the table of values and graph the equation from which it is derived, assuming that it is a straight line:

x	0	1	2	3	4	5
y	$\frac{1}{2}$	$-\frac{3}{4}$				

9. Give the values of m (the slope), $y(0)$, and $y(1)$ for the equation $2x - 5y = -6$.

10. Graph the equations $x = 5$ and $y = -2$.

The answers to this test are on page 282. Check the questions you got wrong on the list below and compare your results with those on Diagnostic Test 5 to see how much you have improved.

1. ______	3. ______	5. ______	7. ______	9. ______
2. ______	4. ______	6. ______	8. ______	10. ______

If you made more than one mistake, review the material and then take the diagnostic test again. Make sure that you get all the questions right before you finish the chapter.

Now proceed to Chap. 6.

ANSWERS TO TESTS FOR CHAPTER 5

DIAGNOSTIC TEST 5

1. $-\frac{3}{16}$

2. -1

3 and 4.

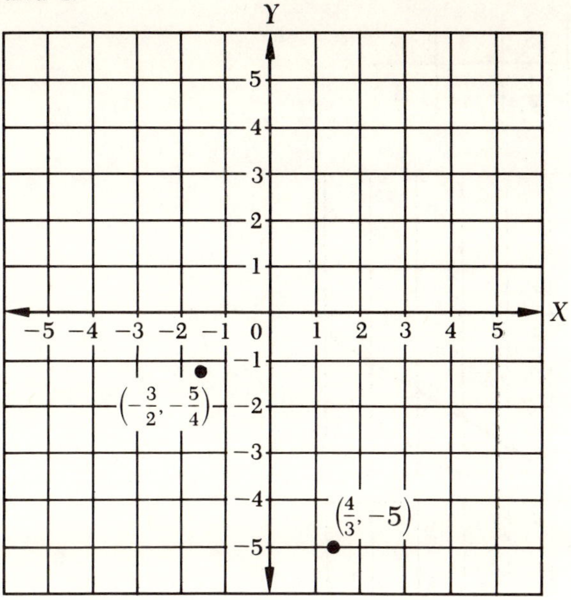

5.

x	0	1	2	3	4	5
y	$-\frac{3}{4}$	-2	$-\frac{13}{4}$	$-\frac{18}{4}$	$-\frac{23}{4}$	-7

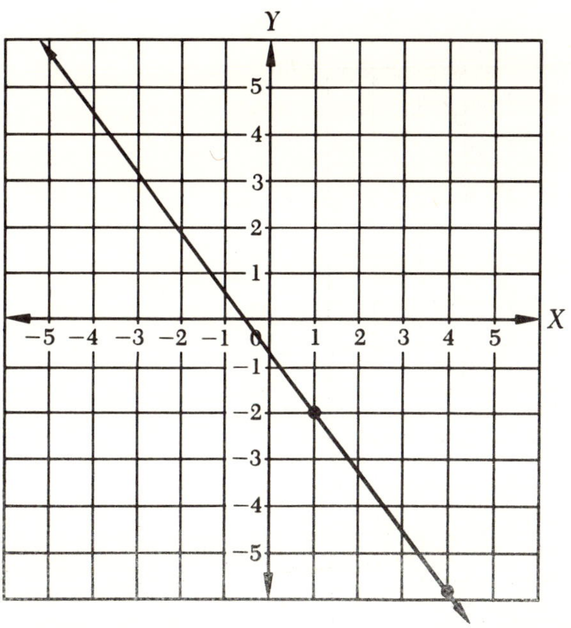

6.

x	0	1	2	3	4	5
y	-1	$-\frac{2}{5}$	$\frac{1}{5}$	$\frac{4}{5}$	$\frac{7}{5}$	2

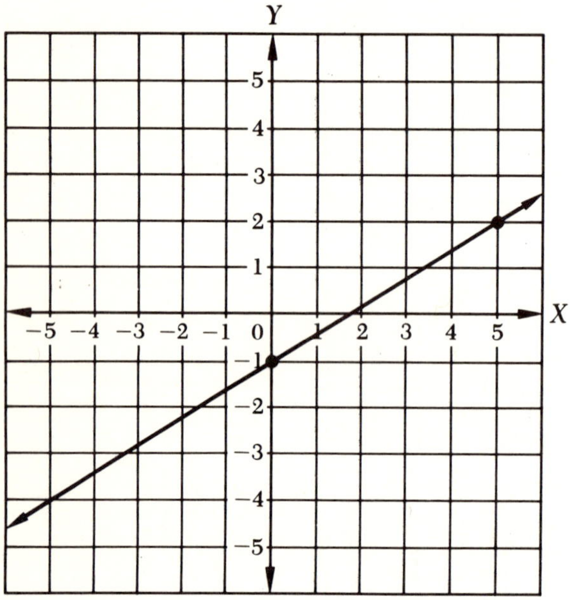

7. $\frac{4}{3}x + \frac{2}{3} = y$; slope $= \frac{4}{3}$, y intercept $= \frac{2}{3}$

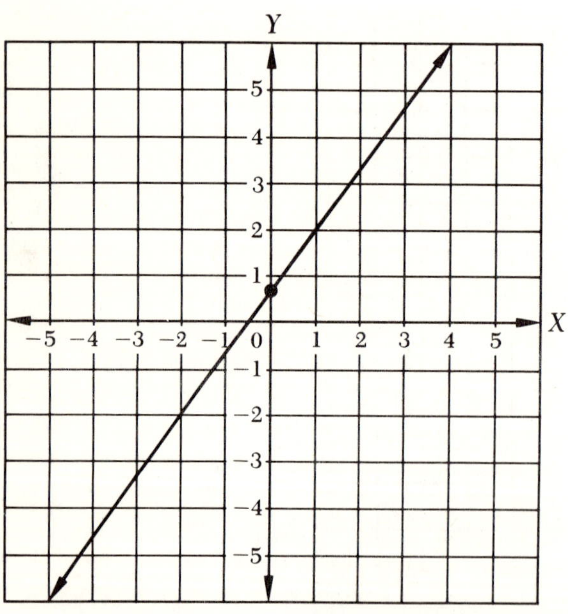

8.

x	0	1	2	3	4	5
y	$-\frac{9}{4}$	$-\frac{3}{2}$	$-\frac{3}{4}$	0	$\frac{3}{4}$	$\frac{3}{2}$

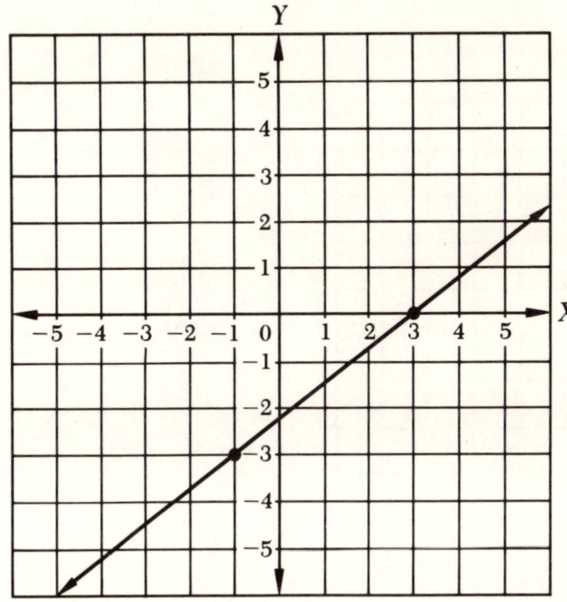

9. $m = -\frac{5}{2}$; $y(0) = -\frac{7}{2}$; $y(1) = -6$

10.

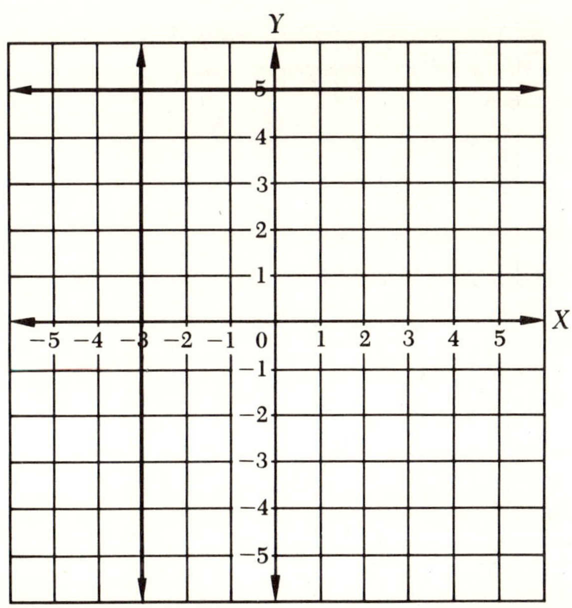

PRETEST 5-1

1. 14
2. $-\frac{11}{3}$
3. $\frac{9}{2}$
4. $-\frac{31}{6}$

POSTTEST 5-1

1. -19
2. $-\frac{11}{4}$
3. $-\frac{1}{5}$
4. $-\frac{32}{9}$

PRETEST 5-2

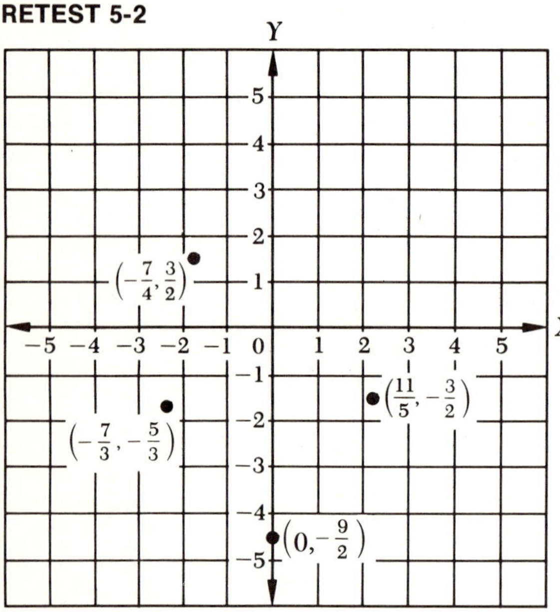

POSTTEST 5-2

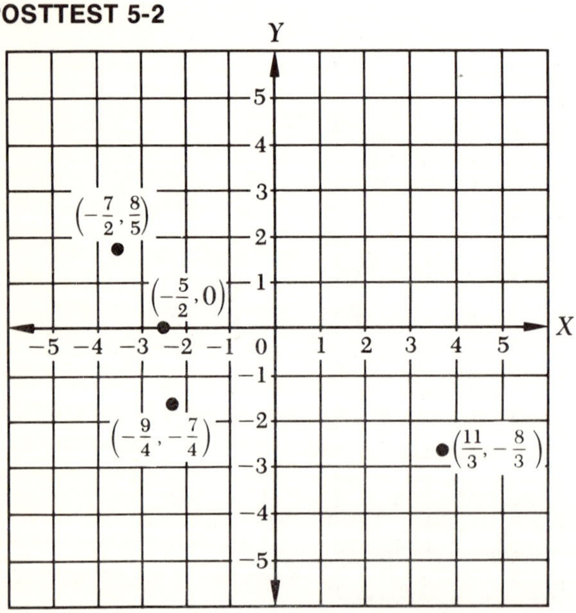

PRETEST 5-3

1.

x	-1	0	1	2	3	4
y	1	$\frac{3}{5}$	$\frac{1}{5}$	$-\frac{1}{5}$	$-\frac{3}{5}$	-1

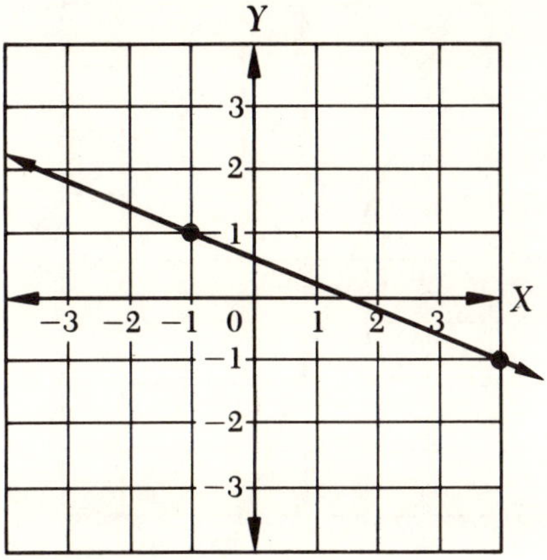

2.

x	-2	-1	0	1	2	3
y	-1	$-\frac{1}{4}$	$\frac{1}{2}$	$\frac{5}{4}$	2	$\frac{11}{4}$

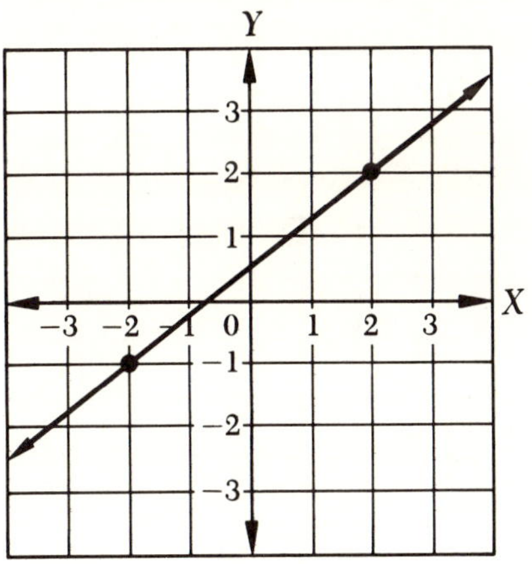

POSTTEST 5-3

1.

x	-2	-1	0	1	2	3
y	2	$\frac{5}{4}$	$\frac{1}{2}$	$-\frac{1}{4}$	-1	$-\frac{7}{4}$

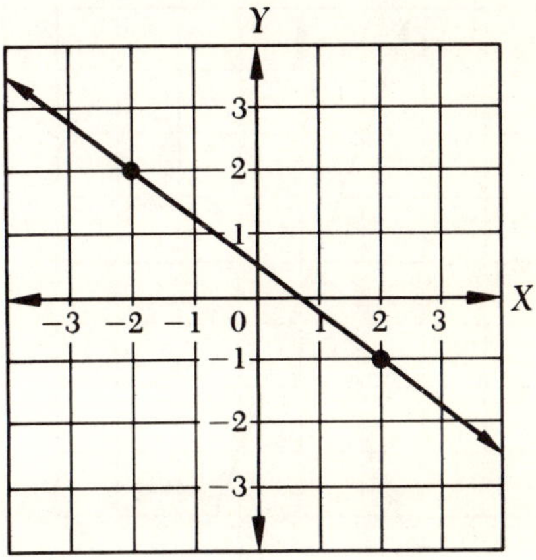

2.

x	-2	-1	0	1	2	3
y	0	$\frac{2}{5}$	$\frac{4}{5}$	$\frac{6}{5}$	$\frac{8}{5}$	2

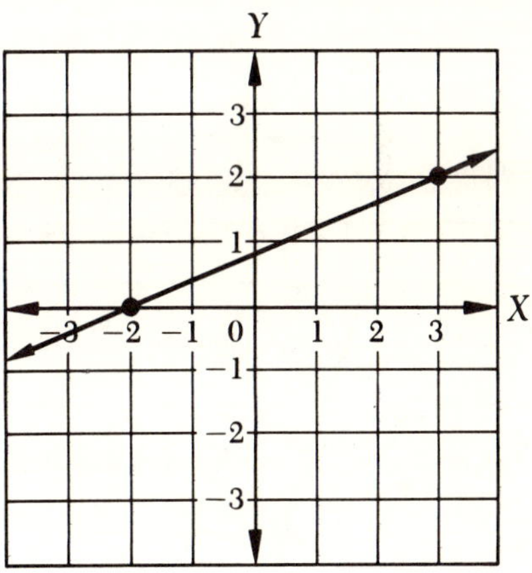

3.

x	-1	0	1	2	3	4	5
y	2	$-\frac{1}{2}$	-3	$-\frac{11}{2}$	-8	$-\frac{21}{2}$	-13

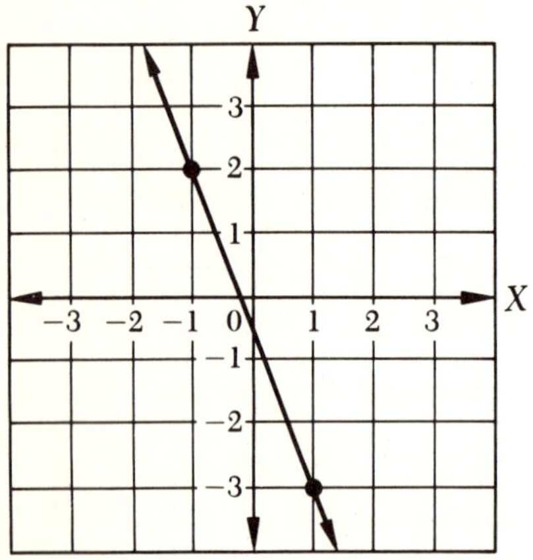

3.

x	-2	-1	0	1	2	3	4
y	3	$\frac{2}{3}$	$-\frac{5}{3}$	-4	$-\frac{19}{3}$	$-\frac{26}{3}$	-11

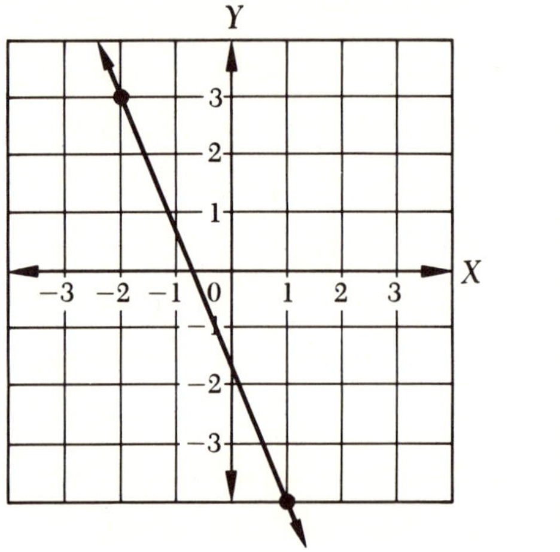

PRETEST 5-4

1. slope $= -\frac{3}{4}$
   Y intercept $= \frac{3}{2}$
2. slope $= \frac{2}{5}$
   Y intercept $= \frac{2}{5}$

3 and 4.

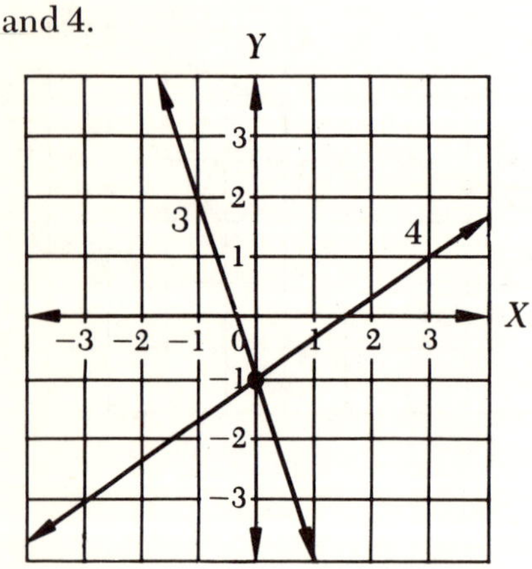

POSTTEST 5-4

1. slope $= -\frac{2}{5}$
   Y intercept $= \frac{6}{5}$
2. slope $= \frac{3}{4}$
   Y intercept $= \frac{1}{2}$

3 and 4.

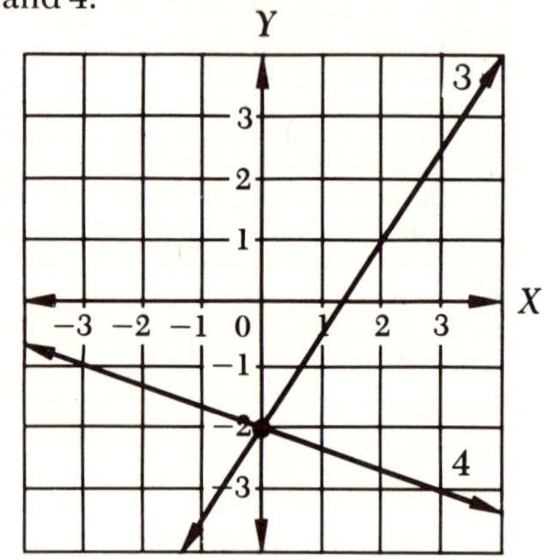

5.

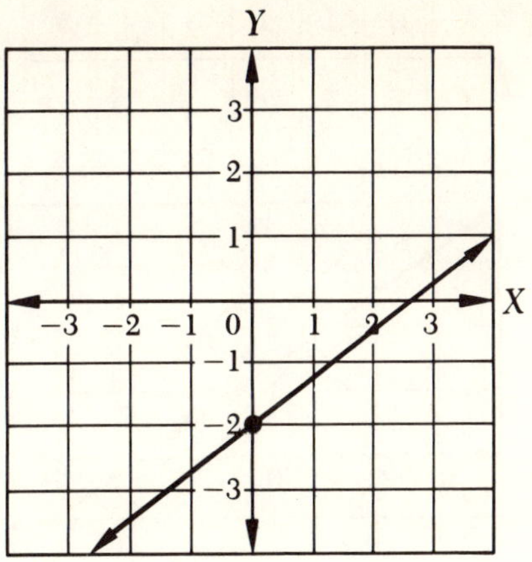

5.

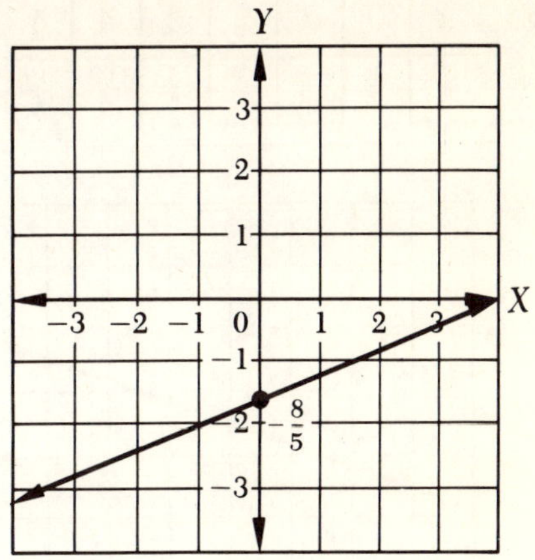

6.

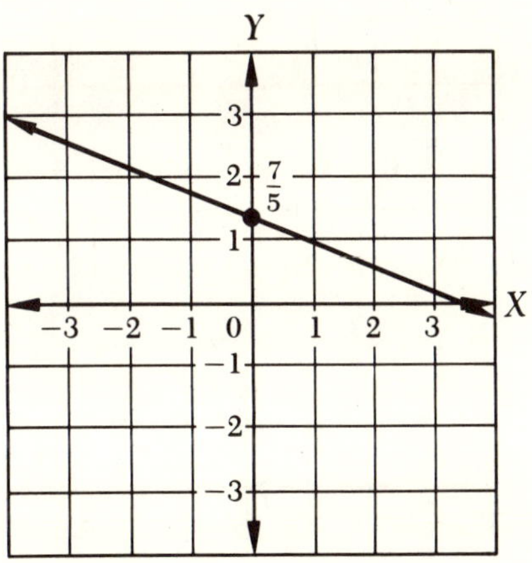

6.

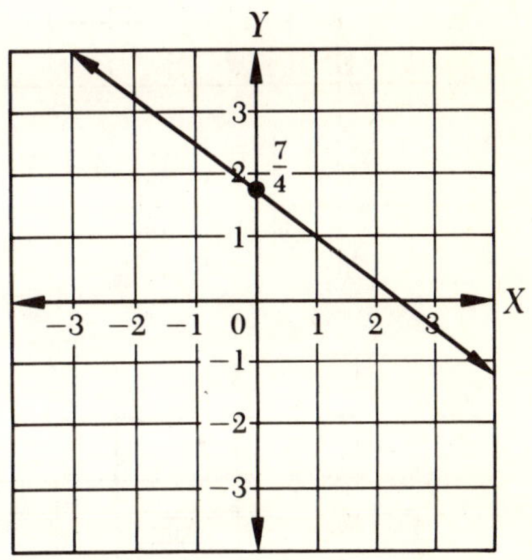

PRETEST 5-5

1. $m = -2$
   $y(0) = -\frac{7}{3}$
   $y(1) = -\frac{13}{3}$
2. $m = \frac{4}{5}$
   $y(0) = \frac{8}{5}$
   $y(1) = \frac{12}{5}$

POSTTEST 5-5

1. $m = \frac{3}{5}$
   $y(0) = \frac{3}{5}$
   $y(1) = \frac{6}{5}$
2. $m = -\frac{1}{2}$
   $y(0) = -\frac{2}{5}$
   $y(1) = -\frac{9}{10}$

3.

x	0	1	2	3	4	5
y	$\frac{1}{2}$	$\frac{5}{6}$	$\frac{7}{6}$	$\frac{3}{2}$	$\frac{11}{6}$	$\frac{13}{6}$

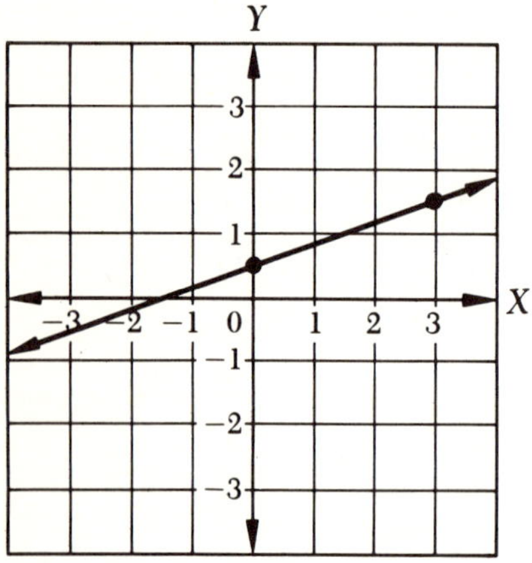

3.

x	0	1	2	3	4	5
y	$\frac{7}{6}$	$\frac{1}{2}$	$-\frac{1}{6}$	$-\frac{5}{6}$	$-\frac{3}{2}$	$-\frac{13}{6}$

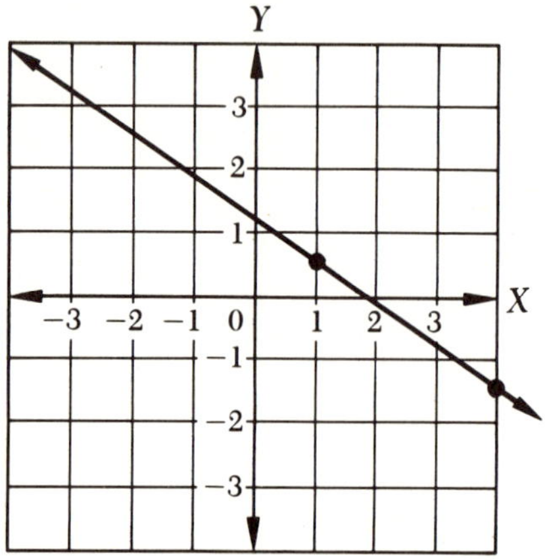

4.

x	0	1	2	3	4	5
y	$-\frac{7}{3}$	$-\frac{11}{3}$	-5	$-\frac{19}{3}$	$-\frac{23}{3}$	-9

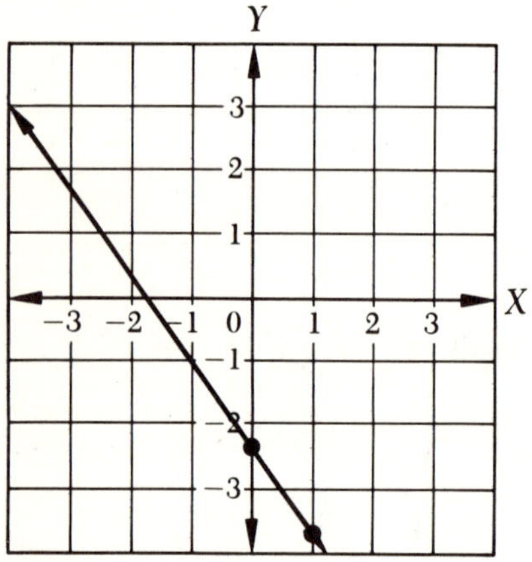

4.

x	0	1	2	3	4	5
y	$-\frac{5}{3}$	$-\frac{1}{3}$	1	$\frac{7}{3}$	$\frac{11}{3}$	5

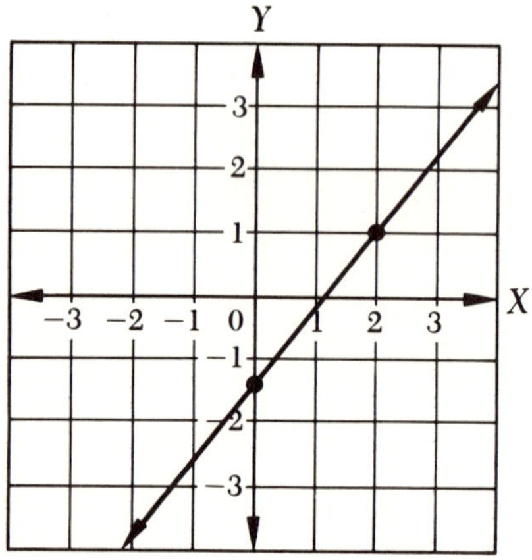

Pretest 5 and 6.

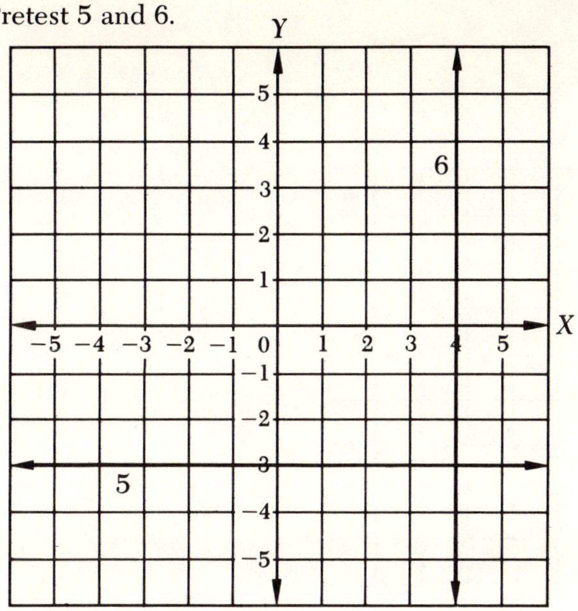

Posttest 5 and 6.

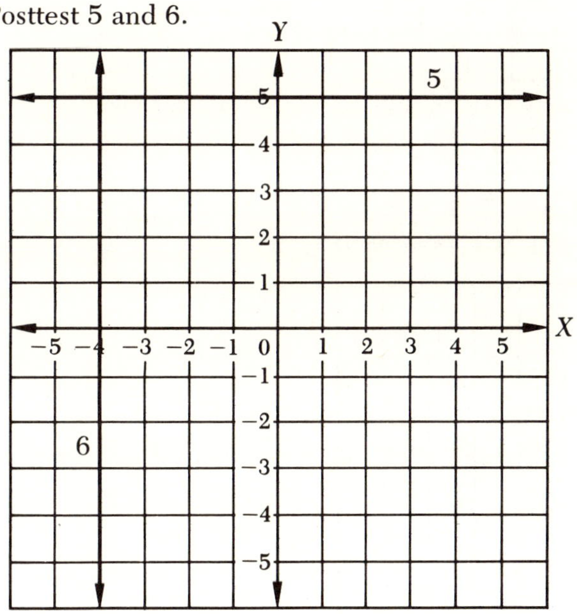

FINAL TEST 5

1. 0

2. $-\frac{33}{2}$

3 and 4.

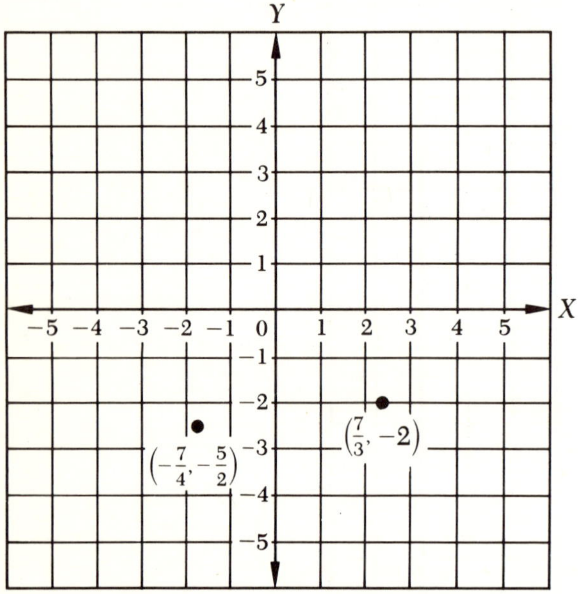

5.

x	0	1	2	3	4	5
y	$-\frac{7}{3}$	-3	$-\frac{11}{3}$	$-\frac{13}{3}$	-5	$-\frac{17}{3}$

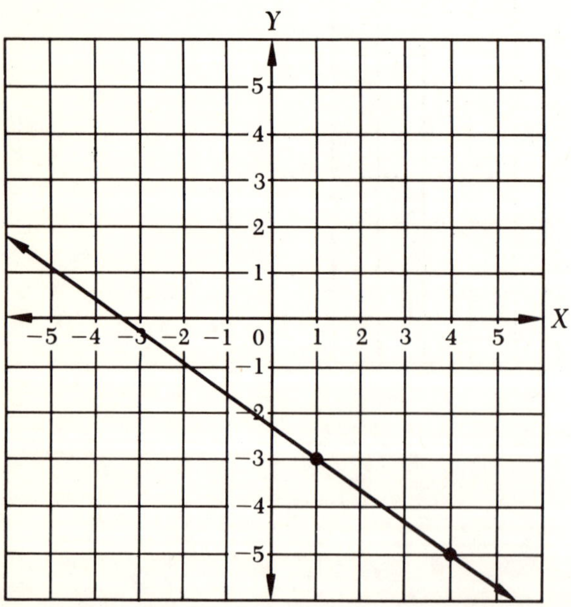

6.

x	0	1	2	3	4	5
y	$-\frac{10}{3}$	-2	$-\frac{2}{3}$	$\frac{2}{3}$	2	$\frac{10}{3}$

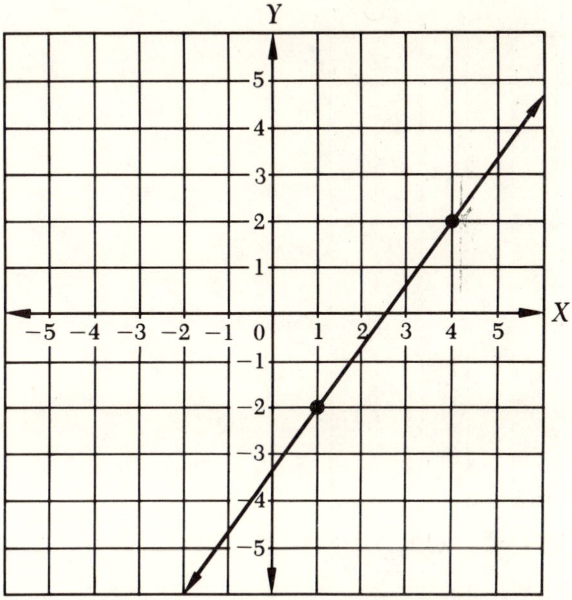

7. $\frac{5}{3}x + \frac{4}{3} = y$; slope is $\frac{5}{3}$, y intercept is $\frac{4}{3}$

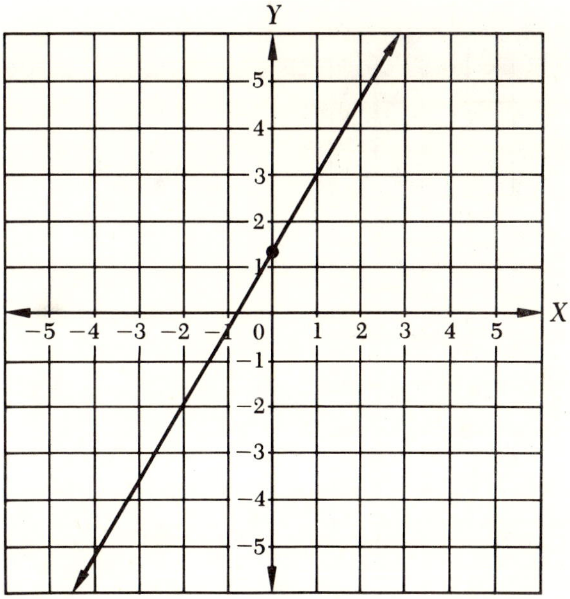

8.

x	0	1	2	3	4	5
y	$\frac{1}{2}$	$-\frac{3}{4}$	-2	$-\frac{13}{4}$	$-\frac{18}{4}$	$-\frac{23}{4}$

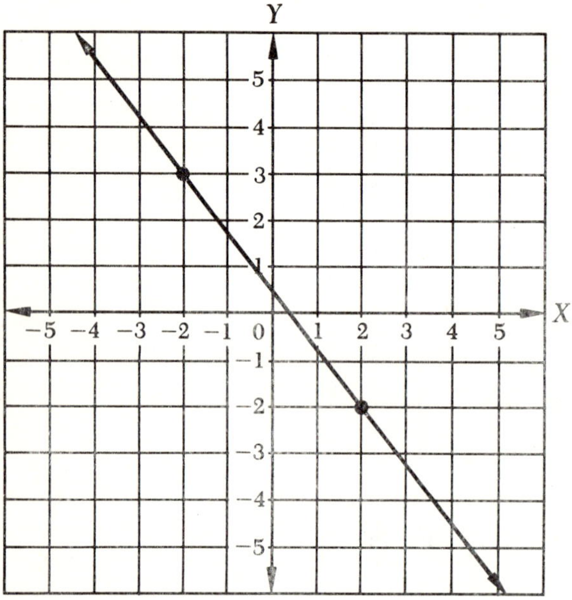

9. $m = \frac{2}{5}$ $\quad y(0) = \frac{6}{5}$ $\quad y(1) = \frac{8}{5}$

10.

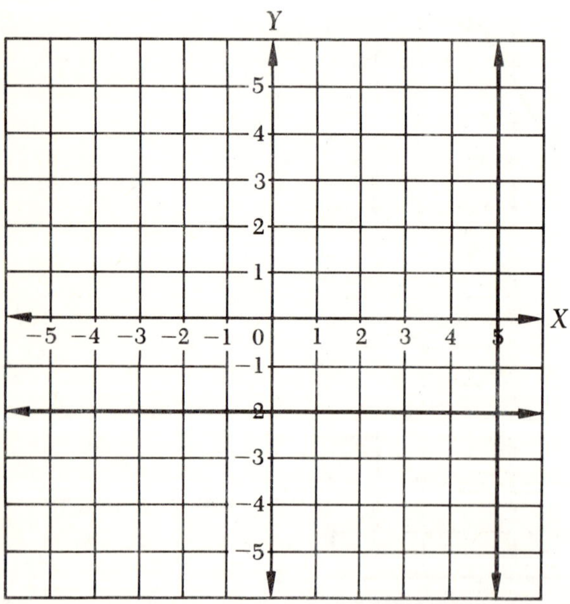

Chapter 6
Pairs of First-degree Equations in Two Variables

This chapter will show you how to find the simultaneous solution sets, i.e., the intersections of the solution sets, of pairs of first-degree equations in two variables. This process is often called *solving simultaneous equations.* Section 6-1 shows you the graphical methods of finding simultaneous solution sets, and simple substitutions such as 3 for y and -2 for x. Section 6-2 extends the substitution method to all pairs of equations, using such substitutions as $\frac{1}{3}x + 5$ for y. Then Sec. 6-3 introduces the addition method and Sec. 6-4 extends it to include equations which contain fractions and parentheses. Sections 6-1, 6-2, and 6-4 all consider those special cases in which the graphs of the two equations concerned are parallel or coincident straight lines.

This chapter requires knowledge of Chaps. 1, 2, and 5. Throughout this chapter the domain of each variable is assumed to be the set of real numbers.

DIAGNOSTIC TEST 6

Give the simultaneous solution sets of the following pairs of equations. Check your answers.

1. $x + 3y = 4$
   $y = -3$

2. $x = 4$
   $x = -4$

3. $y = -2$
   $3y = -6$

4. $y = 5x + 2$
   $3y - 4x = -5$

5. $y = 7 - 3x$
   $6x + 2y = 14$

6. $2x - 4y = -7$
   $8x - 7y = -16$

7. $3x + 8y = 1$
   $5x - 7y = 22$

8. $2x - 5y = 11$
$6x - 2 = 15y$

9. $\dfrac{2x}{3} = \dfrac{5y}{4} + \dfrac{19}{18}$

$\dfrac{7x}{5} + \dfrac{4y}{3} + \dfrac{19}{45} = 0$

10. $\dfrac{2}{5}\left(\dfrac{x}{2} - 3\right) + \dfrac{5}{3}(y + 6) - \dfrac{67}{5} = 0$

$\dfrac{3}{4}\left(\dfrac{x}{9} + 3\right) = \dfrac{2}{7}(y - 4) + \dfrac{199}{84}$

When you have finished the diagnostic test, correct it yourself using the answers on page 330. Then check all the questions you got wrong on the list below:

1. ______	4. ______	9. ______
2. ______	5. ______	10. ______
3. ______	6. ______	
	7. ______	
	8. ______	

If you made no mistakes at all, you need not do any work in this chapter and should go straight on to Chap. 7, First-degree Inequations in Two Variables, which begins on page 333. But if you made any mistakes (even if it was only *one* mistake), follow these instructions:

	section	page
If you made any mistakes in Questions 1 to 3, circle Sec. 6-1.	6-1	287
If you made any mistakes in Questions 4 to 8, circle Secs. 6-2 and 6-3.	6-2 6-3	297 305
If you made any mistakes in Question 9 or 10, circle Sec. 6-4.	6-4	318

You should now work through the sections whose numbers you have circled, taking them in numerical order. The sections begin on the pages indicated.

Section 6-1
The Graphical Method of Finding Simultaneous Solution Sets

Do this section if you made a mistake in Questions 1 to 3 of Diagnostic Test 6. Begin by taking the pretest below, which will show you how much of Sec. 6-1 you already know and what you will be able to do by the time you have finished it.

PRETEST 6-1

1. Graph the three lines given below on the same set of axes:

   A: $x - 2y = 3$ B: $x = 4$ C: $x - 2y = -2$

Give the simultaneous solution sets of the following pairs of lines:

2. A and B 3. A and C 4. B and C

Without graphing give the simultaneous solution sets of the following pairs of lines:

5. $x + 3y = -4$
   $y = -2$
6. $3x - 5y = -2$
   $x = 5$
7. $x = -2$
   $x = 7$
8. $y = 2$
   $4y = 8$

When you have finished the pretest, correct it yourself using the answers on page 331. Then check all the questions you got wrong on the list below:

1. ______	4. ______	7. ______
2. ______	5. ______	8. ______
3. ______	6. ______	

If you made any mistakes (even if it was only *one* mistake), read the introduction and go through the program.

If you made no mistakes at all, skip Sec. 6-1, because you already know its material, and go on to the next section whose number you circled on page 286. If no further numbers were circled, take Final Test 6 (page 329).

Introduction

Section 6-1 introduces the concept of the simultaneous solution set of two equations as the intersection of their solution sets. If the equations are represented by intersecting straight lines, the simultaneous solution set has one element, corresponding to the point of intersection; but if the two lines are parallel, it has no elements; and if the two lines are coincident, it becomes identical with the solution sets of those lines. You will see how the simultaneous solution set of the general equation $ax + by = c$ and equations of the type $x = p$ or $y = q$ can be found by substituting p for x or q for y.

FRAMES FOR SECTION 6-1

1 *a.* What shape is the graph of $2x + 3y = 4$?
b. Is (0,0) a solution of $2x + 3y = 4$?
c. Is $(-1,2)$ a solution of $2x + 3y = 4$?

---

a. a straight line *b.* no: $0 + 0 \neq 4$
c. yes: $2(-1) + 3(2) = -2 + 6 = 4$

---

2 The multiplication transformation also applies to first-degree equations in two variables. Thus

$$x + y = 2$$

$$\updownarrow \text{MT with 3}$$

$$3x + 3y = \underline{\qquad}$$

6

---

3 *a.* Since $x + y = 2$ and $3x + 3y = 6$ are equivalent, they must have (the same/different) ____________ solution set(s).
b. What shape is the graph of $x + y = 2$?
c. What shape is the graph of $3x + 3y = 6$?
d. How is the graph of $x + y = 2$ related to the graph of $3x + 3y = 6$? (Draw the graph if you have any doubts.)

---

a. the same *b.* a straight line *c.* a straight line
d. The graph of $x + y = 2$ is the same straight line as the graph of $3x + 3y = 6$.

---

4 Which of the following equations have the same graph as $2x + y = 5$?

a. $4x + 2y = 10$
b. $-6y - 3y = -15$
c. $8x + 4y = 15$
d. $10x + 5y = 25$
e. $x + \frac{y}{2} = \frac{5}{2}$

---

a, b, d, and *e*

---

5 Two straight lines in the same plane can be related in three possible ways:

a. They can be equivalent, as in ($A/B/C$) ______.
b. They can be parallel, as in ______.
c. They can intersect, as in ______.

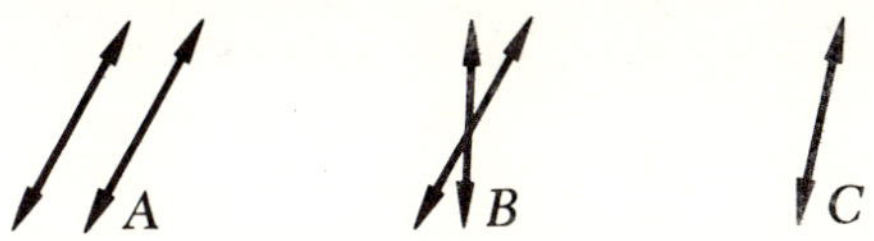

---

a. C *b.* A *c.* B

---

6 *a.* When two lines are equivalent, they have (all/one/none) ______ of their points in common.
b. When two lines are parallel, they have(all/one/none) ______ of their points in common.
c. When two lines intersect, they have (all/one/none) ______ of their points in common.

---

a. all *b.* none *c.* one

---

7 Graph the four lines given below on the same set of axes:

A: $y = 2$ B: $x + y = 2$ C: $x + y = 4$ D: $3y = 6$

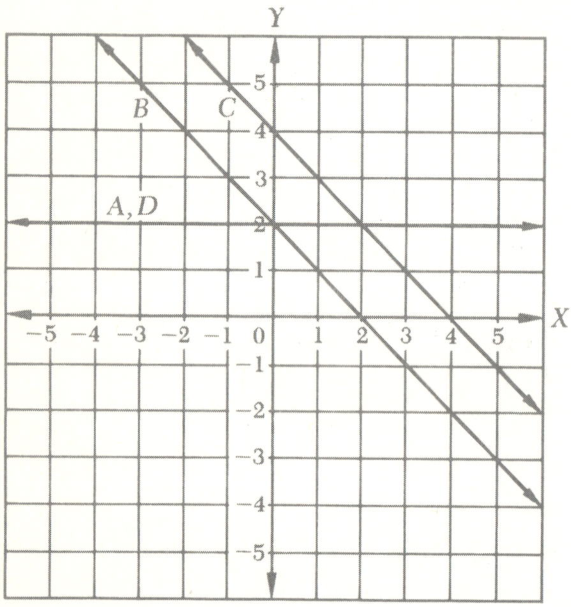

---

8 Look at the answer to Frame 7. Which of the following pairs of lines are coincident, which are parallel, and which intersect? Give the points of intersection where they exist:

a. A and *B* intersect at (0,2).
b. A and *C* ______________.
c. A and *D* ______________.
d. *B* and *C* ______________.
e. *B* and *D* ______________.
f. *C* and *D* ______________.

---

a. intersect at (0,2) *b.* intersect at (2,2) *c.* are coincident
d. are parallel *e.* intersect at (0,2) *f.* intersect at (2,2)

---

9 *a.* When two equations are represented by *parallel* lines, their solution sets have (all/none/one) ______ of their ordered pairs in common.

b. When two equations are represented by *coincident* lines, their solution sets have (all/none/one) ______ of their ordered pairs in common.

c. When two equations are represented by *intersecting* lines, their solution sets have (all/none/one) ______ of their ordered pairs in common.

---

a. none *b.* all *c.* one

---

10 Line A in Frame 7 represents the solution set A of the equation $y = 2$. This set can be described in property notation as $A = \{(x,y) \mid y = 2\}$, because any ordered pair (x,y) for which $y = 2$ is an element of A and is represented by a point on line A. Similarly line B in Frame 7 represents the solution set B of the equation ______; and $B = \{(x,y) \mid$ ______$\}$. Any ordered pair (x,y) for which ______ is an element of B and is represented by a point on line B.

$x + y = 2$ $x + y = 2$ $x + y = 2$

---

11 The lines A and B intersect at the point ______. Is this the only point they have in common? We can write the set of ordered pairs resulting from this intersection as

$$\{(x,y) \mid y = 2\} \cap \{(x,y) \mid x + y = 2\} = \{\_\_\_\_\}$$

or in short as $A \cap B = \{$______$\}$.

(0,2) yes (0,2) (0,2)

---

12 Thus $A \cap B = \{(0,2)\}$ and $A \cap C =$ ______.

$\{(2,2)\}$

---

13 *a.* How many points do lines A and D have in common?
$A \cap D =$ ______ $(A/\emptyset)$.

b. How many points do lines B and C have in common?
$B \cap C =$ ______.

---

a. an infinite number: all their points are in common; A. We could also have written $A \cap D = D$, since A and D are equal *b.* none; $\emptyset$

---

14 Graph the four lines given below on the same set of axes:

E: $4x - 2y = -3$ F: $x = -2$ G: $4x - 2y = 3$ H: $2x = -4$

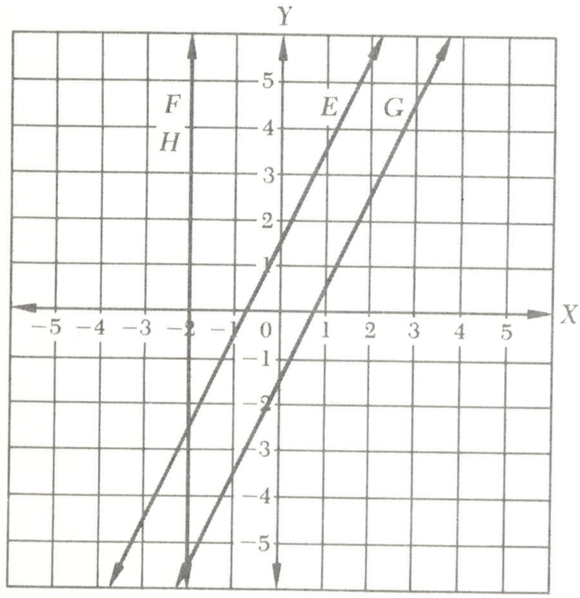

15 *a.* $E \cap F =$ ______   *b.* $E \cap G =$ ______
c. $E \cap H =$ ______   *d.* $F \cap G =$ ______
e. $F \cap H =$ ______   *f.* $G \cap H =$ ______

a. $\{(-2, -\frac{5}{2})\}$ *b.* $\emptyset$ *c.* $\{(-2, -\frac{5}{2})\}$ *d.* $\{(-2, -\frac{11}{2})\}$
e. F or H or $\{(x,y) \mid x = -2\}$ *f.* $\{(-2, -\frac{11}{2})\}$

16 The intersection of the solution sets of two equations is called the *simultaneous solution set* of those two equations, and its elements are called *simultaneous solutions.* What is the simultaneous solution set of $x = -2$ and $4x - 2y = 3$?

$\{(-2, -\frac{11}{2})\}$

17 *a.* How many simultaneous solutions have $4x - 2y = -3$ and $4x - 2y = 3$? What is their simultaneous solution set?

b. How many simultaneous solutions have $x = -2$ and $2x = -4$? What is their simultaneous solution set? (Use property notation.)

a. None; $\emptyset$ *b.* an infinite number; $\{(x,y)\} \mid x = -2\}$

18 Graph the lines $x = 4$, $y = 5$, $y = -1$, and $2x = 8$.

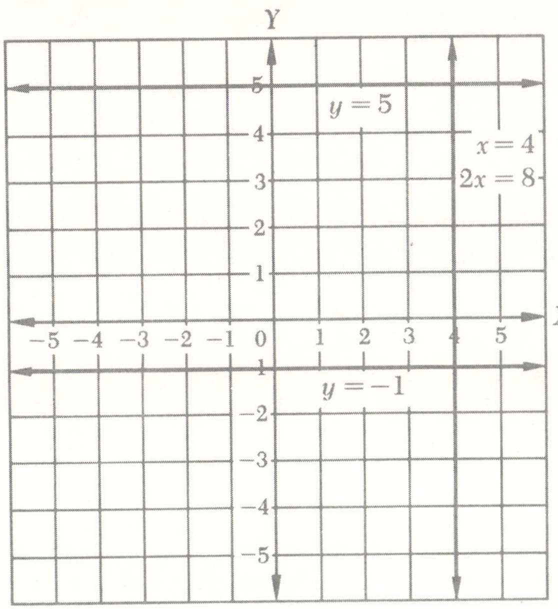

19 Give the simultaneous solution sets of:

a. $x = 4$ and $y = 5$
b. $y = -1$ and $y = 5$
c. $x = 4$ and $2x = 8$
d. $x = 4$ and $y = -1$

a. $\{(4,5)\}$ *b.* $\emptyset$ *c.* $\{(x,y) \mid x = 4\}$ *d.* $\{(4,-1)\}$

20 We can often find the simultaneous solution set of two lines without graphing. For example, to find the simultaneous solution set of $x = 4$ and $x + y = 10$, we can substitute 4 for x in the equation $x + y = 10$. This gives $4 + y = 10 \longleftrightarrow y =$ ______, suggesting that a simultaneous solution is (______,______). Check this answer.

6 (4,6)

CHECK

$4 = 4$ $4 + 6 = 10$

21 *a.* Are there any other solutions of $x + y = 10$ for which $x = 4$? The simultaneous solution set of $x = 4$ and $x + y = 10$ is ________.

b. Use a similar method to find a simultaneous solution of $x = 2$ and $x + y = 10$. Check your answer.

c. What is the simultaneous solution set of $x = 2$ and $x + y = 10$?

---

a. no; $\{(4,6)\}$ *b.* $(2,8)$ CHECK

$2 = 2 \quad 2 + 8 = 10$

c. $\{(2,8)\}$

---

22 Without graphing give the simultaneous solution sets of the following pairs of equations. Check that each of your answers is a solution of both equations.

a. $x + y = 8$, $y = 6$

b. $x - y = 2$, $x = 3$

c. $x - y = 3$, $y = 4$

---

a. $\{(2,6)\}$

CHECK

$2 + 6 = 8 \quad 6 = 6$

b. $\{(3,1)\}$

CHECK

$3 - 1 = 2 \quad 3 = 3$

c. $\{(7,4)\}$

CHECK

$7 - 4 = 3 \quad 4 = 4$

---

23 Without graphing give the simultaneous solution sets of the following pairs of equations. Check that each of your answers is a solution of both equations.

a. $x + y = 2$, $y = -1$

b. $3x + 2y = 5$, $x = -2$

c. $2x - 3y = -10$, $x = -3$

a. $\{(3,-1)\}$

CHECK

$3 + (-1) = 2 \qquad -1 = -1$

b. $\{(-2, \frac{11}{2})\}$

CHECK

$3(-2) + 2(\frac{11}{2}) = -6 + 11 = 5 \qquad -2 = -2$

c. $\{(-3, \frac{4}{3})\}$

CHECK

$2(-3) - 3(\frac{4}{3}) = -6 - 4 = -10 \qquad -3 = -3$

---

24 If we use the same substitution method for the equations $y = 4$ and $y = 5$, we get $4 = 5$, which is (true/false) __________ for all values of y. This indicates that the simultaneous solution set of $y = 4$ and $y = 5$ is __________. What do you know about the graphs of $y = 4$ and $y = 5$?

---

false; $\emptyset$; they are parallel straight lines

---

25 If we use the substitution method for the equations $y = 4$ and $2y = 8$, we get $2(4) = 8$, which is (true/false) ______ for all values of y. This indicates thatthe simultaneous solution set of $y = 4$ and $2y = 8$ is __________. What do you know about the graphs of $y = 4$ and $2y = 8$?

---

true; $\{(x,y) \mid y = 4\}$; they are coincident straight lines

---

26 Give the simultaneous solution sets of:

a. $x + y = 2$
$\quad y = -1$

b. $y = 2$
$\quad y = -1$

c. $2y = -2$
$\quad y = -1$

---

a. $\{(3,-1)\}$ *b.* $\emptyset$ *c.* $\{(x,y) \mid y = -1\}$

You have finished the program for Sec. 6-1. Now take the posttest to see how much you have learned.

POSTTEST 6-1

1. Graph the three lines given below on the same set of axes:

$A: 3x - y = -2 \qquad B: y = 3 \qquad C: 3x - y = 3$

Give the simultaneous solution sets of the following pairs of lines:

2. A and B 3. A and C 4. B and C

Without graphing give the simultaneous solution sets of the following pairs of lines:

5. $2x + 5y = -4$
   $y = 5$

6. $4x - 3y = -3$
   $x = -2$

7. $x = -5$
   $x = 2$

8. $y = 3$
   $3y = 9$

When you have finished the posttest, correct it yourself using the answers on page 331. Then check all the questions you got wrong on the list below:

1. ______ 4. ______ 7. ______

2. ______ 5. ______ 8. ______

3. ______ 6. ______

If you compare your performance with that on Pretest 6-1, you will see how much you have improved. If you made more than one mistake on this posttest, review the material and take the pretest again. Don't move on to the next section until you can get all the questions right.

Now proceed to the next section whose number you circled on page 286. If no other sections were circled, take Final Test 6 (page 329).

Section 6-2
The Substitution Method

Do this section if you made a mistake in Questions 4 to 8 of Diagnostic Test 6. Begin by taking the pretest below, which will show you how much of Sec. 6-2 you already know and what you will be able to do by the time you have finished it.

PRETEST 6-2

Use the method of substitution to find the simultaneous solution sets of the following pairs of equations. Check your answers.

1. $y = 4x - 5$
   $4y - 3x = 6$
2. $2x + 3y = 11$
   $x + 4y = 13$
3. $6x + 7y = -2$
   $3x - 2y = 1$
4. $5x - 2y = 4$
   $2x + 3y = 13$
5. $3y - 2x = 5$
   $4x = 6y - 10$
6. $2y + x = 3$
   $4x = 2 - 8y$

When you have finished the pretest, correct it yourself using the answers on page 332. Then check all the questions you got wrong on the list below:

1. ______ 3. ______ 5. ______

2. ______ 4. ______ 6. ______

If you made any mistakes (even if it was only *one* mistake), read the introduction and go through the program.

If you made no mistakes at all, skip Sec. 6-2, because you already know its material, and go on to Sec. 6-3.

Introduction

You are shown how the method of substitution is extended to solve all pairs of first-degree equations in two variables. This includes the cases for which the two graphs are parallel or coincident lines.

FRAMES FOR SECTION 6-2

1 We can find the simultaneous solution of the equation $x = 2$ and the equation $3x + 2y = 12$ by substituting $x = 2$ in the second equation. This gives us the sentence ______ $+ 2y = 12 \longleftrightarrow y =$ ______.

---

6 3

---

2 We can use the same substitution procedure to find the simultaneous solution of the equations $x = y + 2$ and $3x + 2y = 12$.
Substituting $x = y + 2$ in the second equation, we get

$$3(y + 2) + 2y = 12$$
$$3y + \_\_\_\_\_\_ + 2y = 12$$
$$5y = \_\_\_\_\_\_$$
$$y = \_\_\_\_\_\_$$

Since $x = y + 2$, $x =$ ______.

---

6 6 $\frac{6}{5}$ $\frac{16}{5}$

---

3 Is $(\frac{16}{5}, \frac{6}{5})$ a solution of $x = y + 2$? Is $(\frac{16}{5}, \frac{6}{5})$ a solution of $3x + 2y = 12$? What is the simultaneous solution set of $x = y + 2$ and $3x + 2y = 12$?

---

yes; yes; $\{(\frac{16}{5}, \frac{6}{5})\}$

---

4 Perhaps we should explain why this substitution procedure gives the right answer. Suppose that $\{(a,b)\}$ is the simultaneous solution set of $x = y + 4$ and $2x + 5y = 15$. Is (a,b) a solution of $x = y + 4$? ______ Is $a = b + 4$ a true statement? ______ Is (a,b) a solution of $2x + 5y = 15$? ______ Is $2a + 5b = 15$ a true statement? ______

---

yes; yes; yes; yes

---

5 The fact that $a = b + 4$ and $2a + 5b = 15$ are true statements gives us useful information about a and b. For example, a can be replaced by $b + 4$ since they are equal to each other. This gives us two new statements: $b + 4 = b + 4$ and ______ $+ 5b = 15$.

$2(b + 4)$ or $2b + 8$

---

6 $b + 4 = b + 4$ is clearly true for all values of b, but $2(b + 4) + 5b = 15$ will only be true if $2b +$ _____ $+ 5b = 15$, or if $7b =$ _____, or if $b =$ _____, and if $b =$ _____, we know that $a = b + 4 =$ _____.

---

8 7 1 1 5

---

7 Thus the statements $a = b + 4$ and $2a + 5b = 15$ cannot both be true unless $a = 5$ and $b = 1$. Thus the simultaneous solution set of $x = y + 4$ and $2x + 5y = 15$ is _____.

$\{(5, 1)\}$

---

8 What is the simultaneous solution set of $x = 2y + 8$ and $3x + 4y = 4$? Show your work and check your answer.

$$\begin{aligned} 3(2y + 8) + 4y &= 4 \\ 6y + 24 + 4y &= 4 \\ 10y &= -20 \\ y &= -2 \\ x &= 2(-2) + 8 = -4 + 8 = 4 \end{aligned}$$

Simultaneous solution set is $\{(4, -2)\}$.

CHECK

$3(4) + 4(-2) = 12 + (-8) = 4$

---

9 What is the simultaneous solution set of $y = 3x - 2$ and $2x + 3y = -28$? Show your work and check your answer.

$$\begin{aligned} 2x + 3(3x - 2) &= -28 \\ 2x + 9x - 6 &= -28 \\ 11x &= -22 \\ x &= -2 \\ y &= 3(-2) - 2 = -6 - 2 = -8 \end{aligned}$$

Simultaneous solution set is $\{(-2, -8)\}$.

CHECK

$2(-2) + 3(-8) = (-4) + (-24) = (-28)$

---

10 Often we have to convert an equation to the form $y = nx + b$ or $x = my + c$ before we can use the method of substitution. Thus

$$2x + 3y = 7$$
$$2x = 7 - 3y$$
$$x = \frac{7}{2} - \frac{3y}{2}$$

or

$$2x + 3y = 7$$
$$3y = \_\_\_\_\_$$
$$y = \_\_\_\_\_$$

$7 - 2x \qquad \frac{7}{3} - \frac{2x}{3}$

---

11 *a.* Convert the equation $4x + 5y = 9$ to the form $x =$ ____________.
b. Convert the equation $3x - 7y = 9$ to the form $y =$ ____________.

---

a. $4x + 5y = 9$
$$4x = 9 - 5y$$
$$x = \frac{9}{4} - \frac{5y}{4}$$

b. $3x - 7y = 9$
$$3x - 9 = 7y$$
$$y = \frac{3x}{7} - \frac{9}{7}$$

---

12 Use the method of substitution to find the simultaneous solution set of

$$2x + 3y = 7 \qquad \text{and} \qquad 4x - 6y = 11$$

Show your work and check your answer with both equations.

Method 1

$$x = \frac{7}{2} - \frac{3y}{2}$$

$$4\left(\frac{7}{2} - \frac{3y}{2}\right) - 6y = 11$$

$$14 - 6y - 6y = 11$$

$$-12y = -3$$

$$y = \frac{1}{4}$$

$$x = \frac{7}{2} - \frac{3}{2}\left(\frac{1}{4}\right)$$

$$= \frac{7}{2} - \frac{3}{8}$$

$$= \frac{28 - 3}{8}$$

$$= \frac{25}{8}$$

Method 2

$$y = \frac{7}{3} - \frac{2x}{3}$$

$$4x - 6\left(\frac{7}{3} - \frac{2x}{3}\right) = 11$$

$$4x - 14 + 4x = 11$$

$$8x = 25$$

$$x = \frac{25}{8}$$

$$y = \frac{7}{3} - \frac{2}{3}\left(\frac{25}{8}\right)$$

$$= \frac{7}{3} - \frac{25}{12}$$

$$= \frac{28 - 25}{12}$$

$$= \frac{3}{12} = \frac{1}{4}$$

Simultaneous solution set is $\left\{\left(\frac{25}{8}, \frac{1}{4}\right)\right\}$.

CHECK

$$4\left(\frac{25}{8}\right) - 6\left(\frac{1}{4}\right) = \frac{25}{2} - \frac{3}{2} = \frac{22}{2} = 11$$

$$2\left(\frac{25}{8}\right) + 3\left(\frac{1}{4}\right) = \frac{25}{4} + \frac{3}{4} = \frac{28}{4} = 7$$

13 Use the method of substitution to find the simultaneous solution set of

$$4x - 7y = -8 \qquad \text{and} \qquad 6x + 8y = 25$$

Check your answer with both equations.

$\left\{\left(\frac{3}{2}, 2\right)\right\}$

CHECK

$$4\left(\frac{3}{2}\right) - 7(2) = 6 - 14 = -8$$

$$6\left(\frac{3}{2}\right) + 8(2) = 9 + 16 = 25$$

14 If we use the method of substitution to find the simultaneous solution of $2x + 3y = 7$ and $6y = 15 - 4x$, then we substitute

$$\frac{15}{6} - \frac{4x}{6} \quad \text{or} \quad \frac{5}{2} - \frac{2x}{3} \quad \text{for } y \text{ in the first equation}$$

a. This gives

$$2x + 3\left(\frac{5}{2} - \frac{2x}{3}\right) = 7$$

$$2x + \_\_\_\_ - \_\_\_\_ = 7$$

$$\_\_\_\_ = 7$$

which is (true/false) ____ for all values of x and y.

b. The simultaneous solution set of $2x + 3y = 7$ and $6y = 15 - 4x$ is ____.

c. What can you say about the graphs of these two equations?

---

a. $\frac{15}{2} - 2x$; $\frac{15}{2}$; false *b.* $\emptyset$ *c.* They are parallel straight lines.

---

15 If we use the method of substitution to find the simultaneous solution of $2x + 8y = 4$ and $2 - 4y = x$, we substitute $2 - 4y$ for x in the first equation.

a. This gives

$$2(2 - 4y) + 8y = 4$$

$$\_\_\_\_ - \_\_\_\_ + \_\_\_\_ = 4$$

$$\_\_\_\_ = 4$$

which is (true/false) ____ for all values of x and y.

b. The simultaneous solution set of $2x + 8y = 4$ and $4y - 2 = x$ is ____.

c. What can you say about the graphs of these two equations?

---

a. $4 - 8y + 8y$ 4 true

b. $\{(x,y) \mid x = 2 - 4y\}$ or $\{(x,y) \mid 2x + 8y = 4\}$

c. They are coincident straight lines.

---

16 Use the method of substitution to find the simultaneous solution set of:

a. $2x - 3y = 4$ and $5 + 6y = 4x$

b. $2x - 3y = 4$ and $8 + 6y = 4x$

$$2x - 3y = 4 \longleftrightarrow x = 2 + \frac{3y}{2}$$

Substituting for x in the second equation,

a. $5 + 6y = 4\left(2 + \frac{3y}{2}\right) = 8 + 6y \longleftrightarrow 5 = 8$

Simultaneous solution set is $\emptyset$.

b. $8 + 6y = 4\left(2 + \frac{3y}{2}\right) = 8 + 6y$

Simultaneous solution set is $\{(x,y) \mid 2x - 3y = 4\}$ or $\{(x,y) \mid 8 + 6y = 4x\}$.

17 Use the method of substitution to find the simultaneous solution set of $5x - 4y = -3$ and $2x + 9y = 7$. Check your answer with both equations.

$$\left\{\left(\frac{1}{53}, \frac{41}{53}\right)\right\}$$

CHECK

$$5\left(\frac{1}{53}\right) - 4\left(\frac{41}{53}\right) = \frac{5}{53} - \frac{164}{53} = -\frac{159}{53} = -3$$

$$2\left(\frac{1}{53}\right) + 9\left(\frac{41}{53}\right) = \frac{2}{53} + \frac{369}{53} = \frac{371}{53} = 7$$

18 Use the method of substitution to find the simultaneous solution set of $3x - 5y = -7$ and $7x + 2y = -3$. Check your answer with both equations.

$$\left\{\left(-\frac{29}{41}, \frac{40}{41}\right)\right\}$$

CHECK

$$3\left(-\frac{29}{41}\right) - 5\left(\frac{40}{41}\right) = -\frac{87}{41} - \frac{200}{41} = -\frac{287}{41} = -7$$

$$7\left(-\frac{29}{41}\right) + 2\left(\frac{40}{41}\right) = -\frac{203}{41} + \frac{80}{41} = -\frac{123}{41} = -3$$

You have finished the program for Sec. 6-2. Now take the posttest to see how much you have learned.

POSTTEST 6-2

Use the method of substitution to find the simultaneous solution sets of the following pairs of equations. Check your answers.

1. $x = 3y + 5$
   $4x + 3y = 5$
2. $3x + 2y = -10$
   $x - 3y = 4$
3. $4x - 5y = -7$
   $5x + 10y = 6$
4. $4y - 3x = 6$
   $2x + 3y = 16$
5. $5x - 3y = 7$
   $6y = 11 + 10x$
6. $x - 4y = 3$
   $3x = 12y + 9$

When you have finished the posttest, correct it yourself using the answers on page 332. Then check all the questions you got wrong on the list below:

1. ______ 3. ______ 5. ______

2. ______ 4. ______ 6. ______

If you compare your performance with that on Pretest 6-2, you will see how much you have improved. If you made more than one mistake on this posttest, review the material and take the pretest again. Don't move on to the next section until you can get all the questions right.

Section 6-3
The Addition Method

Do this section if you made a mistake in Questions 4 to 8 of Diagnostic Test 6. Begin by taking the pretest below, which will show you how much of Sec. 6-3 you already know and what you will be able to do by the time you have finished it.

PRETEST 6-3

Use the method of addition to find the simultaneous solution sets of the following pairs of equations. Check your answers.

1. $5x + 2y = 3$
   $7x - 2y = 9$
2. $4x + 5y = 7$
   $2x - 3y = -13$
3. $3x - 5y = -7$
   $7x - 2y = 3$
4. $7x - 6y = 2$
   $3x + 16y = 12$

When you have finished the pretest, correct it yourself using the answers on page 332. Then check all the questions you got wrong on the list below:

1. ______ 3. ______

2. ______ 4. ______

If you made any mistakes (even if it was only *one* mistake), read the introduction and go through the program.

If you made no mistakes at all, skip Sec. 6-3, because you already know its material, and go on to the next section whose number you circled on page 286. If no further numbers were circled, take Final Test 6 (page 329).

Introduction

You will learn how to substitute for multiples of x and y. You will also see that a substitution for nx can be equivalent to the addition of two equations which both have nx as a term. After the validity of the method of addition has been demonstrated, you are shown how it is extended to the solution of all pairs of equations by the conversion of pairs of equations to forms which have a common term in x or y.

FRAMES FOR SECTION 6-3

1 In order to find the simultaneous solution of equations

$$3x = 9 - 4y \qquad \text{and} \qquad 3x + 8y = 1$$

it is easiest to substitute for $3x$ rather than for x. Thus

$$3x + 8y = 1$$

becomes

$$\begin{aligned}(9 - 4y) + 8y &= 1\\ 9 + 4y &= 1\\ 4y &= \_\_\_\_\\ y &= \_\_\_\_\end{aligned}$$

and

$$\begin{aligned}3x &= 9 - 4y = 9 + \_\_\_\_ = \_\_\_\_\\ x &= \_\_\_\_\end{aligned}$$

The simultaneous solution set is ________. Check your answer with both equations.

$$4y = -8$$

$$y = -2$$

$$3x = 9 + 8 = 17$$

$$x = \frac{17}{3}$$

$$\left\{\left(\frac{17}{3}, -2\right)\right\}$$

CHECK

$$3\left(\frac{17}{3}\right) = 9 - 4(-2) = 9 + 8$$

$$3\left(\frac{17}{3}\right) + 8(-2) = 17 - 16 = 1$$

---

2 Find the simultaneous solution set of $5x = 7y - 4$ and $5x + 3y = 16$ by substituting for $5x$. Show your work and check your answer.

$$(7y - 4) + 3y = 16$$
$$10y - 4 = 16$$
$$10y = 20$$
$$y = 2$$
$$5x = 7(2) - 4 = 10$$
$$x = 2$$

Simultaneous solution set is $\{(2,2)\}$

CHECK

$$5(2) = 7(2) - 4 = 10$$
$$5(2) + 3(2) = 16$$

3 Find the simultaneous solution set of $3x + 7y = -8$ and $5x - 7y = -24$ by substituting for $7y$. Show your work and check your answer.

Method 1

$$7y = -8 - 3x$$
$$5x - (-8 - 3x) = -24$$
$$5x + 8 + 3x = -24$$
$$8x = -32$$
$$x = -4$$
$$7y = -8 - 3(-4)$$
$$= -8 + 12 = 4$$
$$y = \frac{4}{7}$$

Method 2

$$7y = 5x + 24$$
$$3x + (5x + 24) = -8$$
$$8x + 24 = -8$$
$$8x = -32$$
$$x = -4$$
$$7y = 5(-4) + 24$$
$$= -20 + 24 = 4$$
$$y = \frac{4}{7}$$

Simultaneous solution set is $\left\{\left(-4, \frac{4}{7}\right)\right\}$.

CHECK

$$3(-4) + 7\left(\frac{4}{7}\right) = -12 + 4 = -8$$
$$5(-4) - 7\left(\frac{4}{7}\right) = -20 - 4 = -24$$

4 This procedure can be shortened still further by adding the two equations:

$$\begin{array}{rr} \text{Adding} & 3x + 7y = -8 \\ \text{to} & 5x - 7y = -24 \\ \hline \text{gives} & 3x + 5x + 7y - 7y = -8 - 24 \\ & 8x = \_\_\_\_ \end{array}$$

Notice that this (is/is not) ____ the same equation as the one we obtained when we substituted for $7y$.

-32 is

---

5 It is no coincidence that we obtained the same equation:

a. Add

$$\begin{array}{rr} \text{Add} & mx + ny = a \\ \text{to} & px - ny = b \\ \hline \end{array}$$

$$mx + px \_\_\_\_\_\_ = \_\_\_\_\_\_$$

$$mx + px = \_\_\_\_\_\_$$

b. If we substitute $ny = px - b$ in the top equation, we get

$$mx + (px - b) = a$$

$$mx \_\_\_\_\_\_ = a$$

$$mx + px = \_\_\_\_\_\_$$

---

a. $mx + px + ny - ny = a + b$
$mx + px = a + b$

b. $mx + px - b = a$
$mx + px = a + b$

---

6 Nor is it so strange that the addition of equations is valid. If $x = 4$, $y = 2$, then

a. $x + y =$ ____ *b.* $x + 2y =$ ____

c. $2x + 3y =$ ____ *d.* $3x + 5y =$ ____

e. What do we get by adding Eq. *a* to Eq. *b*?

f. What do we get by adding Eq. *b* to Eq. *c*?

g. What equation do we get by adding Eq. *c* to Eq. *d*?

h. Which of these equations have (4,2) as a solution?

---

a. 6 *b.* 8 *c.* 14 *d.* 22

e. $2x + 3y = 14$ *f.* $3x + 5y = 22$ *g.* $5x + 8y = 36$

h. all of them

---

7 Use the method of adding equations to obtain the simultaneous solution set of

$2x + 9y = 4$ and $5x - 9y = 10$

Show your work and check your answer.

$$\begin{aligned} 2x + 9y &= 4 \\ 5x - 9y &= 10 \\ \hline 7x &= 14 \\ x &= 2 \\ 2(2) + 9y &= 4 \\ 9y &= 4 - 4 = 0 \\ y &= 0 \end{aligned}$$

Simultaneous solution set is $\{(2,0)\}$.

CHECK

$2(2) + 0 = 4$
$5(2) - 0 = 10$

---

8 Use the method of adding equations to obtain the simultaneous solution set of

$4y - 3x = 5$ and $3x + 2y = 7$

Show your work and check your answer.

$$\begin{aligned} -3x + 4y &= 5 \\ 3x + 2y &= 7 \\ \hline 6y &= 12 \\ y &= 2 \\ 3x + 2(2) &= 7 \\ 3x &= 7 - 4 = 3 \\ x &= 1 \end{aligned}$$

Simultaneous solution set is $\{(1,2)\}$.

$4(2) - 3(1) = 8 - 3 = 5$
$3(1) + 2(2) = 3 + 4 = 7$

---

9 The method of adding is most useful when one of the variables is eliminated. In Frame 7, adding the two equations eliminates the variable (x/y) ______, because one equation contains the term ______ and the other equation contains the term ______ and the sum of these two terms is ______. In Frame 8, the variable (x/y) ______ is eliminated because one equation contains the term ______ and the other equation contains the term ______ and the sum of these two terms is ______.

---

y $9y, -9y$ (either order) 0 x $-3x, 3x$ (either order) 0

---

10 It is often best to replace one or both of the equations by an equivalent equation before adding. To find the simultaneous solution set of $5x + 6y = 4$ and $7x - 2y = 3$, you multiply the second equation by 3 before adding (this gives $-6y$, the negative of $6y$, and $-6y + 6y = 0$):

$$7x - 2y = 3 \xleftrightarrow[3]{\text{MT with}} \begin{array}{r} 21x - 6y = 9 \\ +5x + 6y = 4 \\ \hline \underline{\qquad} = \underline{\quad} \\ x = \underline{\quad} \end{array}$$

Find y and then check your answer.

$$26x = 13$$

$$x = \frac{1}{2}$$

$$7\left(\frac{1}{2}\right) - 2y = 3$$

$$7 - 4y = 6$$

$$4y = 1$$

$$y = \frac{1}{4}$$

Simultaneous solution set is $\left\{\left(\frac{1}{2}, \frac{1}{4}\right)\right\}$.

CHECK

$$5\left(\frac{1}{2}\right) + 6\left(\frac{1}{4}\right) = \frac{5}{2} + \frac{3}{2} = \frac{8}{2} = 4$$

$$7\left(\frac{1}{2}\right) - 2\left(\frac{1}{4}\right) = \frac{7}{2} - \frac{1}{2} = \frac{6}{2} = 3$$

---

11 Find the simultaneous solution set of $7x + 8y = 3$ and $2x - 4y = 15$.

$$2x - 4y = 15 \overset{\text{MT with}}{\underset{2}{\longleftrightarrow}} \begin{array}{rcl} \underline{\qquad\qquad} & = & \underline{\quad} \\ +7x + 8y & = & 3 \\ \hline \end{array}$$

Finish the problem and check your answer.

$$2x - 4y = 15 \longleftrightarrow \begin{array}{rcl} 4x - 8y & = & 30 \\ 7x + 8y & = & 3 \\ \hline 11x & = & 33 \\ x & = & 3 \end{array}$$

$$\begin{aligned} 2(3) - 4y &= 15 \\ 6 - 4y &= 15 \\ 4y &= -9 \\ y &= -\frac{9}{4} \end{aligned}$$

Simultaneous solution set is $\left\{\left(3, -\frac{9}{4}\right)\right\}$.

CHECK

$$7(3) + 8\left(-\frac{9}{4}\right) = 21 - 18 = 3$$

$$2(3) - 4\left(-\frac{9}{4}\right) = 6 + 9 = 15$$

---

12 Find the simultaneous solution set of $2x + 7y = 3$ and $10x + 3y = -1$.

$$2x + 7y = 3 \xleftrightarrow[-5]{\text{MT with}} \begin{array}{r} \_\_\_\_\_\_\_\_\_\_ = \_\_\_ \\ +10x + 3y = -1 \\ \hline \end{array}$$

Finish the problem and check your answer.

$$2x + 7y = 3 \longleftrightarrow \begin{array}{r} -10x - 35y = -15 \\ 10x + 3y = -\ 1 \\ \hline -32y = -16 \\ y = \frac{1}{2} \end{array}$$

$$2x + 7\left(\frac{1}{2}\right) = 3$$

$$2x + \frac{7}{2} = 3$$

$$4x + 7 = 6$$

$$4x = -1$$

$$x = -\frac{1}{4}$$

Simultaneous solution set is $\left\{\left(-\frac{1}{4}, \frac{1}{2}\right)\right\}$.

CHECK

$$2\left(-\frac{1}{4}\right) + 7\left(\frac{1}{2}\right) = -\frac{1}{2} + \frac{7}{2} = \frac{6}{2} = 3$$

$$10\left(-\frac{1}{4}\right) + 3\left(\frac{1}{2}\right) = -\frac{5}{2} + \frac{3}{2} = -\frac{2}{2} = -1$$

---

13 Find the simultaneous solution set of $5x + 2y = -4$ and $7x + 6y = 4$. Check your answer.

$$5x + 2y = -4 \xleftrightarrow[-3]{\text{MT with}} \begin{array}{rcr} -15x - 6y &=& 12 \\ 7x + 6y &=& 4 \\ \hline -8x &=& 16 \\ x &=& -2 \end{array}$$

$$\begin{aligned} 5(-2) + 2y &= -4 \\ -10 + 2y &= -4 \\ 2y &= 6 \\ y &= 3 \end{aligned}$$

Simultaneous solution set is $\{(-2,3)\}$

CHECK

$$5(-2) + 2(3) = -10 + 6 = -4$$
$$7(-2) + 6(3) = -14 + 18 = 4$$

---

14 Sometimes both equations have to be replaced by equivalent equations before adding will eliminate a variable. To find the simultaneous solution set of $5x - 2y = 12$ and $4x + 3y = 5$

$$5x - 2y = 12 \xleftrightarrow[\text{MT with } 3]{} 15x - 6y = 36$$

$$4x + 3y = 5 \xleftrightarrow[\text{MT with } 2]{} \underline{8x + 6y = 10}$$

Finish the problem and check your answer.

$$\begin{aligned} 23x &= 46 \\ x &= 2 \\ 5(2) - 2y &= 12 \\ 10 - 2y &= 12 \\ 2y &= -2 \\ y &= -1 \end{aligned}$$

Simultaneous solution set is $\{(2,-1)\}$.

CHECK

$$5(2) - 2(-1) = 10 + 2 = 12$$
$$4(2) + 3(-1) = 8 - 3 = 5$$

15 Find the simultaneous solution set of $4x + 11y = 2$ and $7x + 19y = 3$. Check your answer.

$$4x + 11y = 2 \overset{\text{MT}}{\underset{\text{with } 7}{\longleftrightarrow}} \_\_\_\_\_\_ = \_\_\_$$

$$7x + 19y = 3 \overset{\text{MT}}{\underset{\text{with } -4}{\longleftrightarrow}} \_\_\_\_\_\_ = \_\_\_$$

$$\begin{aligned} 4x + 11y = 2 &\longleftrightarrow \quad 28x + 77y = 14 \\ 7x + 19y = 3 &\longleftrightarrow \quad -28x - 76y = -12 \\ & \qquad\qquad\quad y = 2 \end{aligned}$$

$$\begin{aligned} 4x + 11(2) &= 2 \\ 4x + 22 &= 2 \\ 4x &= -20 \\ x &= -5 \end{aligned}$$

Simultaneous solution set is $\{(-5, 2)\}$.

CHECK

$4(-5) + 11(2) = -20 + 22 = 2$
$7(-5) + 19(2) = -35 + 38 = 3$

---

16 Look at Frame 14. The coefficients of y are ___ in the first equation and ___ in the second equation. In order to eliminate the variable y we multiplied the first equation by ___ and the second equation by ___. Look at Frame 15. The coefficients of x are ___ in the first equation and ___ in the second equation. In order to eliminate the variable x we multiplied the first equation by ___ and the second equation by ___.

-2 3 3 2 4 7 7 -4

---

17 Find the simultaneous solution set of $3x + 7y = -12$ and $4x + 5y = -3$. Check your answer.

$$3x + 7y = -12 \overset{\text{MT}}{\underset{\text{with } 4}{\longleftrightarrow}} \_\_\_\_\_\_ = \_\_\_$$

$$4x + 5y = -3 \overset{\text{MT}}{\underset{\text{with } \_}{\longleftrightarrow}} \_\_\_\_\_\_ = \_\_\_$$

$$3x + 7y = -12 \xleftrightarrow[\text{with } 4]{\text{MT}} 12x + 28y = -48$$

$$4x + 5y = -3 \xleftrightarrow[\text{with } -3]{\text{MT}} -12x - 15y = 9$$

$$13y = -39$$

$$y = -3$$

$$\begin{aligned} 3x + 7(-3) &= -12 \\ 3x - 21 &= -12 \\ 3x &= 9 \\ x &= 3 \end{aligned}$$

Simultaneous solution set is $\{(3, -3)\}$.

CHECK

$$3(3) + 7(-3) = 9 - 21 = -12$$
$$4(3) + 5(-3) = 12 - 15 = -3$$

18 Find the simultaneous solution set of $14x - 5y = 23$ and $8x + 3y = 19$. Check your answer.

$$14x - 5y = 23 \xleftrightarrow[\text{with } 3]{\text{MT}} \underline{\qquad\qquad} = \underline{\qquad}$$

$$8x + 3y = 19 \xleftrightarrow[\text{with } \_]{\text{MT}} \underline{\qquad\qquad} = \underline{\qquad}$$

$$\underline{\qquad\qquad\qquad\qquad}$$

$$14x - 5y = 23 \xleftrightarrow[\text{with } 3]{\text{MT}} 42x - 15y = 69$$

$$8x + 3y = 19 \xleftrightarrow[\text{with } 5]{\text{MT}} 40x + 15y = 95$$

$$82x = 164$$

$$x = 2$$

$$\begin{aligned} 14(2) - 5y &= 23 \\ 28 - 5y &= 23 \\ 5y &= 5 \\ y &= 1 \end{aligned}$$

Simultaneous solution set is $\{(2, 1)\}$.

CHECK

$$14(2) - 5(1) = 28 - 5 = 23$$
$$8(2) + 3(1) = 16 + 3 = 19$$

19 Find the simultaneous solution set of $11x + 19y = -16$ and $2x - 7y = 18$. Check your answer. (It doesn't matter which variable you choose to eliminate. It is best to keep the numbers as small as possible.)

$$11x + 19y = -16 \xleftrightarrow[\text{with } 2]{\text{MT}} 22x + 38y = -32$$

$$2x - 7y = 18 \xleftrightarrow[\text{with } -11]{\text{MT}} -22x + 77y = -198$$

$$115y = -230$$

$$y = -2$$

$$\begin{aligned} 11x + 19(-2) &= -16 \\ 11x - 38 &= -16 \\ 11x &= 22 \\ x &= 2 \end{aligned}$$

Simultaneous solution set is $\{(2,-2)\}$.

CHECK

$$11(2) + 19(-2) = 22 - 38 = -16$$
$$2(2) - 7(-2) = 4 + 14 = 18$$

---

20 Find the simultaneous solution set of $12x + 5y = 1$ and $9x - 4y = -38$. Check your answer.

$\{(-2,5)\}$

CHECK

$$12(-2) + 5(5) = -24 + 25 = 1$$
$$9(-2) - 4(5) = -18 - 20 = -38$$

---

21 Find the simultaneous solution set of $2x - 13y = 57$ and $7x - 11y = 27$. Check your answer.

$\{(-4,-5)\}$

CHECK

$$2(-4) - 13(-5) = -8 + 65 = 57$$
$$7(-4) - 11(-5) = -28 + 55 = 27$$

---

22 Find the simultaneous solution set of $4x - 9y = 15$ and $7x + 8y = -7$. Check your answer.

$$\left\{\left(\frac{3}{5}, -\frac{7}{5}\right)\right\}$$

CHECK

$$7\left(\frac{3}{5}\right) + 8\left(-\frac{7}{5}\right) = \frac{21}{5} - \frac{56}{5} = -\frac{35}{5} = -7$$

$$4\left(\frac{3}{5}\right) - 9\left(-\frac{7}{5}\right) = \frac{12}{5} + \frac{63}{5} = \frac{75}{5} = 15$$

You have finished the program for Sec. 6-3. Now take the posttest to see how much you have learned.

POSTTEST 6-3

Use the method of addition to find the simultaneous solution sets of the following pairs of equations. Check your answers.

1. $3x + 4y = 10$
   $5x - 4y = 6$
2. $4x - 7y = 16$
   $5y - 12x = -16$
3. $3x + 8y = -18$
   $2x - 9y = 31$
4. $9x - 5y = -2$
   $7x + 9y = 21$

When you have finished the posttest, correct it yourself using the answers on page 332. Then check all the questions you got wrong on the list below.

1. ______ 3. ______

2. ______ 4. ______

If you compare your performance with that on Pretest 6-3, you will see how much you have improved. If you made more than one mistake on this posttest, review the material and take the pretest again. Don't move on to the next section until you can get all the questions right.

Now proceed to the next section whose number you circled on page 286. If no other sections were circled, take Final Test 6 (page 329).

Section 6-4
Equations Containing Parentheses and Fractions

Do this section if you made a mistake in Question 9 or 10 of Diagnostic Test 6. Begin by taking the pretest below, which will show you how much of Sec. 6-4 you already know and what you will be able to do by the time you have finished it.

PRETEST 6-4

Find the simultaneous solution sets of the following equations. Check your answers.

1. $2x - 3y = 5$
   $9y = 6x - 15$

2. $3x + 5y + 9 = 0$
   $2y = 7x - 20$

3. $4x = 2y + 3$
   $4y = 8x - 7$

4. $\dfrac{4x}{3} + \dfrac{3y}{4} = 7$
   $\dfrac{3x}{2} - \dfrac{5y}{4} = -\dfrac{1}{2}$

5. $\dfrac{2x}{5} = \dfrac{5y}{7} - \dfrac{3}{35}$
   $\dfrac{7x}{3} + \dfrac{8y}{9} + \dfrac{50}{9} = 0$

6. $\dfrac{3}{4}\left(\dfrac{x}{2} - 3\right) + \dfrac{2}{5}(y - 4) + \dfrac{31}{4} = 0$
   $\dfrac{5}{3}(x + 6) = \dfrac{2}{7}\left(\dfrac{y}{9} + 3\right) + \dfrac{8}{3}$

When you have finished the pretest, correct it yourself using the answers on page 332. Then check all the questions you got wrong on the list below:

1. ______ 3. ______ 5. ______

2. ______ 4. ______ 6. ______

If you made any mistakes (even if it was only *one* mistake), read the introduction and go through the program.

If you made no mistakes at all, skip Sec. 6-4, because you already know its material, and go on to take Final Test 6 (page 329).

Introduction

You will learn how equations can be converted to a common form ($ax + by = c$; $by = c - ax$; etc.) before addition, and you will see that this process immediately detects parallel and coincident lines. You are then shown how to extend the methods of the previous sections to find the simultaneous solution sets of equations containing fractions and parentheses.

FRAMES FOR SECTION 6-4

1 First-degree equations in two variables can be written in a number of different forms. In the last section we used the form $ax + by = c$. $ax + by = c$ can be converted to other forms by using the addition transformation.

a. $ax + by = c$

↕ Adding $-c$

$ax + by - c =$ ______

b. $ax + by = c$

↕ Adding $-ax$

$by =$ ______

c. $ax + by = c$

↕ Adding $-by$

$ax =$ ______

---

a. 0 *b.* $c - ax$ *c.* $c - by$

---

2 We now have four forms of the same equation:

1. $ax + by = c$
2. $ax + by - c = 0$
3. $by = c - ax$
4. $ax = c - by$

Which of the following equations is in which form?

a. $8y = 6 + 5x$
b. $10x + 11y + 2 = 0$
c. $11x = -4 + 8y$
d. $4x - 7y = -2$

---

a. 3 *b.* 2 *c.* 4 *d.* 1

---

3 Convert the equations $8y = 6 + 5x$ and $10x + 11y + 2 = 0$ to the same form as $11x = -4 + 8y$

$-5x = 6 - 8y \qquad 10x = -2 - 11y$

---

4 Convert the equations $8y = 6 + 5x$ and $10x + 11y + 2 = 0$ to the same form as $4x - 7y = -2$.

$-5x + 8y = 6 \qquad 10x + 11y = -2$

---

5 Finding the simultaneous solution set of two equations by the addition method is easiest when both equations are in the same form:

$$3x - 4y = 5$$
$$8x = 9 + 2y$$

a. Put the bottom equation in the same form as the top one.

b. Find the simultaneous solution set and check your answer in the *original* equations.

---

a. $8x - 2y = 9$

b. $8x - 2y = 9 \xleftrightarrow[\text{with } -2]{\text{MT}}$

$$\begin{aligned} -16x + 4y &= -18 \\ 3x - 4y &= 5 \\ \hline -13x &= -13 \\ x &= 1 \end{aligned}$$

$$\begin{aligned} 3(1) - 4y &= 5 \\ 3 &= 5 + 4y \\ -2 &= 4y \\ y &= -\frac{1}{2} \end{aligned}$$

Simultaneous solution set is $\left\{\left(1, -\frac{1}{2}\right)\right\}$.

CHECK

$$3x - 4y = 3(1) - 4\left(-\frac{1}{2}\right) = 3 + 2 = 5$$

$$9 + 2y = 9 + 2\left(-\frac{1}{2}\right) = 9 - 1 = 8 = 8x$$

---

6 $7x = 3 + 5y$
$2x + 7 = 3y$

a. Put the top equation in the same form as the bottom one.
b. Find the simultaneous solution set and check your answer in the *original* equations.

---

a. $7x - 3 = 5y$

b. $7x - 3 = 5y \xleftrightarrow[\text{with } 2]{\text{MT}} 14x - 6 = 10y$

$2x + 7 = 3y \xleftrightarrow[\text{with } -7]{\text{MT}} -14x - 49 = -21y$

$$-55 = -11y$$
$$5 = y$$

$2x + 7 = 3(5) = 15$
$2x = 8$
$x = 4$

Simultaneous solution set is $\{(4,5)\}$.

CHECK

$3 + 5y = 3 + 5(5) = 28 = 7x$
$2x + 7 = 2(4) + 7 = 15 = 3y$

---

7 $2x + 9y = 12$
$4y = 5x + 23$

Find the simultaneous solution set by the method of addition and check your answer in the *original* equations.

$\{(-3,2)\}$

CHECK

$2x + 9y = 2(-3) + 9(2) = -6 + 18 = 12$
$5x + 23 = 5(-3) + 23 = -15 + 23 = 8 = 4y$

---

8 $4x = -12 - 7y$
$4y = 5x - 2$

Find the simultaneous solution set by the method of addition and check your answer in the *original* equations.

$\left\{\left(-\frac{2}{3}, -\frac{4}{3}\right)\right\}$

CHECK

$$-12 - 7y = -12 + \frac{28}{3} = \frac{-36 + 28}{3} = -\frac{8}{3} = 4x$$

$$5x - 2 = -\frac{10}{3} - 2 = \frac{-10 - 6}{3} = -\frac{16}{3} = 4y$$

---

9 If the simultaneous solution set is empty or infinite, this becomes apparent as soon as the two equations are prepared for addition. Thus, if we have $5x - 3y = 7$ and $6y = 11 + 10x$ and convert the second equation to the form of the first, we have

$$10x - 6y = -11$$

↕ MT with $\frac{1}{2}$

$$5x - 3y = -\frac{11}{2}$$

Clearly no ordered pair can satisfy $5x - 3y = 7$ and $5x - 3y = -\frac{11}{2}$, and so their simultaneous solution set is ______.

$\emptyset$

---

10 With $x - 4y = 3$ and $3x = 12y + 9$, converting the first to the form of the second gives

$$x = 4y + 3$$

↕ MT with 3

$$3x = 12y + 9$$

Since the two equations are equivalent, their simultaneous solution set is the same as their solution set, that is, ________

$\{(x,y) \mid x = 4y + 3\}$

11 Find the simultaneous solution sets of

a. $2x - 7y = 9$
$6x = 21y + 27$

b. $4x - y = 6$
$3y = 7 + 12x$

a. $\{(x,y) \mid 2x - 7y = 9\}$ *b.* $\emptyset$

12 $3x + 4y + 7 = 0$
$12y = 5x + 14$

Find the simultaneous solution set by the method of addition and check your answer in the *original* equations.

$\left\{\left(-\frac{5}{2}, \frac{1}{8}\right)\right\}$

CHECK

$$3x + 4y + 7 = 3\left(-\frac{5}{2}\right) + 4\left(\frac{1}{8}\right) + 7 = -\frac{15}{2} + \frac{1}{2} + 7 = 0$$

$$5x + 14 = 5\left(-\frac{5}{2}\right) + 14 = \frac{-25 + 28}{2} = \frac{3}{2} = 12y$$

13 With equations containing fractions, it is easier to remove fractions before finding the simultaneous solution in the usual way. To find the simultaneous solution set of

$$\frac{3}{2}x + \frac{2}{3}y + \frac{1}{6} = 0 \qquad \text{and} \qquad \frac{3}{5}x = \frac{2}{7}y - \frac{41}{35}$$

a. $\frac{3}{2}x + \frac{2}{3}y + \frac{1}{6} = 0$

$\updownarrow$ MT with 6

$9x + \_y + \_ = 0$

b. $\frac{3}{5}x = \frac{2}{7}y - \frac{41}{35}$

$\updownarrow$ MT with 35

$21x = \_y - \_$

c. By the usual methods, simultaneous solution set is ______.

d. Check this answer in the original equations.

---

a. 4; 1 *b.* 10; 41 *c.* $\{(-1,2)\}$

d. CHECK

$$\frac{3}{2}x + \frac{2}{3}y + \frac{1}{6} = \frac{3}{2}(-1) + \frac{2}{3}(2) + \frac{1}{6}$$

$$= -\frac{3}{2} + \frac{4}{3} + \frac{1}{6} = \frac{-9+8+1}{6} = 0$$

$$\frac{2}{7}y - \frac{41}{35} = \frac{2}{7}(2) - \frac{41}{35} = \frac{4}{7} - \frac{41}{35} = \frac{20-41}{35} = \frac{-21}{35}$$

$$= -\frac{3}{5} = \frac{3}{5}x$$

---

14 To find the simultaneous solution set of

$$\frac{3}{4}y = \frac{2}{3}x - 5 \qquad \text{and} \qquad \frac{7}{12}x = \frac{5}{16}y + 3$$

a. $\frac{3}{4}y = \frac{2}{3}x - 5$

$\updownarrow$ MT with 12

$\_\_y = \_\_\_\_\_\_\_\_\_\_\_\_$

b. $\frac{7}{12}x = \frac{5}{16}y + 3$

$\updownarrow$ MT with 48

$\_\_x = \_\_\_\_\_\_\_\_\_\_\_\_$

c. Simultaneous solution set is ______.

d. Check this answer in the original equations.

---

a. $9y = 8x - 60$ *b.* $28x = 15y + 144$ *c.* $\{(3,-4)\}$

d. CHECK

$$\frac{2}{3}x - 5 = \frac{2}{3}(3) - 5 = 2 - 5 = -3 = \frac{3}{4}y$$

$$\frac{5}{16}y + 3 = \frac{5}{16}(-4) + 3 = -\frac{5}{4} + 3 = \frac{-5 + 12}{4} = \frac{7}{4} = \frac{7}{12}x$$

---

15 Find the simultaneous solution set of

$$\frac{5x}{8} - \frac{3y}{5} = -\frac{11}{20} \qquad \text{and} \qquad \frac{3x}{4} = \frac{2y}{7} + \frac{9}{14}$$

Check your answer in the original equations.

$\{(2,3)\}$

CHECK

$$\frac{5x}{8} - \frac{3y}{5} = \frac{5(2)}{8} - \frac{3(3)}{5} = \frac{5}{4} - \frac{9}{5} = \frac{25 - 36}{20} = -\frac{11}{20}$$

$$\frac{2y}{7} + \frac{9}{14} = \frac{6}{7} + \frac{9}{14} = \frac{12 + 9}{14} = \frac{21}{14} = \frac{3}{2} = \frac{3x}{4}$$

---

16 When an equation contains parentheses as well as fractions, it is easier if you first remove the parentheses and then remove the fractions and simplify. To find the simultaneous solution set of

$$\frac{1}{2}\left(\frac{x}{3} + 2\right) + \frac{1}{3}\left(\frac{y}{4} + 7\right) = \frac{41}{12} \quad \text{and} \quad \frac{1}{5}\left(\frac{y}{2} + 6\right) = \frac{1}{3}\left(\frac{x}{5} + \frac{1}{2}\right) + \frac{5}{3}$$

a.

$$\frac{1}{2}\left(\frac{x}{3} + 2\right) + \frac{1}{3}\left(\frac{y}{4} + 7\right) = \frac{41}{12}$$

$\updownarrow$ Remove parentheses

$$\_\_\_\_ + \_\_\_\_ + \_\_\_\_ + \_\_\_\_ = \frac{41}{12}$$

$\updownarrow$ MT with 12

$$\_\_\_\_ + \_\_\_\_ + \_\_\_\_ + \_\_\_\_ = \_\_\_\_$$

$\updownarrow$ Simplify

$$\_\_\_\_x + \_\_\_\_y = \_\_\_\_$$

b.

$$\frac{1}{5}\left(\frac{y}{2} + 6\right) = \frac{1}{3}\left(\frac{x}{5} + \frac{1}{2}\right) + \frac{5}{3}$$

$\updownarrow$ Remove parentheses

$$\_\_\_\_ + \_\_\_\_ = \_\_\_\_ + \_\_\_\_ + \frac{5}{3}$$

$\updownarrow$ MT with 30

$$\_\_\_\_ + \_\_\_\_ = \_\_\_\_ + \_\_\_\_ + \_\_\_\_$$

$\updownarrow$ Simplify

$$\_\_\_\_x + \_\_\_\_y = \_\_\_\_$$

c. Simultaneous solution set is ________.

d. Check your answer in the original equations.

---

a. $\frac{x}{6} + 1 + \frac{y}{12} + \frac{7}{3} = \frac{41}{12}$

$2x + 12 + y + 28 = 41$

$2x + y = 1$

b. $\frac{y}{10} + \frac{6}{5} = \frac{x}{15} + \frac{1}{6} + \frac{5}{3}$

$3y + 36 = 2x + 5 + 50$

$-2x + 3y = 19$

c. $\{(-2,5)\}$

d. CHECK

$$\frac{1}{2}\left(-\frac{2}{3} + 2\right) + \frac{1}{3}\left(\frac{5}{4} + 7\right) = \frac{1}{2}\left(\frac{4}{3}\right) + \frac{1}{3}\left(\frac{33}{4}\right)$$

$$= \frac{2}{3} + \frac{11}{4} = \frac{8 + 33}{12} = \frac{41}{12}$$

$$\frac{1}{3}\left(\frac{x}{5} + \frac{1}{2}\right) + \frac{5}{3} = \frac{1}{3}\left(\frac{-2}{5} + \frac{1}{2}\right) + \frac{5}{3} = \frac{1}{3}\left(-\frac{4}{10} + \frac{5}{10}\right) + \frac{5}{3}$$

$$= \frac{1}{30} + \frac{50}{30} = \frac{51}{30} = \frac{17}{10}$$

$$\frac{1}{5}\left(\frac{y}{2} + 6\right) = \frac{1}{5}\left(\frac{5}{2} + 6\right) = \frac{1}{5}\left(\frac{17}{2}\right) = \frac{17}{10}$$

---

17 Find the simultaneous solution set of

$$\frac{2}{3}\left(\frac{x}{4} + 3\right) = \frac{2}{5}(y + 7) - \frac{1}{6} \quad \text{and} \quad \frac{5}{2}\left(\frac{x}{3} + 3\right) - \frac{2}{3}(y - 4) - \frac{32}{3} = 0$$

and check your answer in the original equations.

$\{(-1, -2)\}$

CHECK

$$\frac{2}{5}(y + 7) - \frac{1}{6} = \frac{2}{5}(-2 + 7) - \frac{1}{6} = \frac{2}{5}(5) - \frac{1}{6} = 2 - \frac{1}{6} = \frac{11}{6}$$

$$\frac{2}{3}\left(\frac{x}{4} + 3\right) = \frac{2}{3}\left(-\frac{1}{4} + 3\right) = \frac{2}{3}\left(\frac{-1 + 12}{4}\right) = \frac{2}{3}\left(\frac{11}{4}\right) = \frac{11}{6}$$

$$\frac{5}{2}\left(-\frac{1}{3} + 3\right) - \frac{2}{3}(-2 - 4) - \frac{32}{3} = \frac{5}{2}\left(\frac{8}{3}\right) - \frac{2}{3}(-6) - \frac{32}{3}$$

$$= \frac{20}{3} + 4 - \frac{32}{3} = 4 - \frac{12}{3} = 0$$

---

18 Find the simultaneous solution set of

$$\frac{1}{4}\left(\frac{x}{7}+6\right)-\frac{1}{7}\left(\frac{3y}{4}+\frac{3}{2}\right)-\frac{29}{28}=0 \qquad \text{and}$$

$$\frac{1}{2}\left(\frac{2y}{5}+3\right)=\frac{2}{3}\left(\frac{4x}{5}-\frac{1}{2}\right)+\frac{41}{30}$$

and check your answer in the original equations.

$\{(2,3)\}$

CHECK

$$\frac{1}{4}\left(\frac{2}{7}+6\right)-\frac{1}{7}\left(\frac{9}{4}+\frac{3}{2}\right)-\frac{29}{28}=\frac{1}{4}\left(\frac{44}{7}\right)-\frac{1}{7}\left(\frac{9+6}{4}\right)-\frac{29}{28}$$

$$=\frac{11}{7}-\frac{15}{28}-\frac{29}{28}=\frac{44-15-29}{28}=0$$

$$\frac{2}{3}\left(\frac{4x}{5}-\frac{1}{2}\right)+\frac{41}{30}=\frac{2}{3}\left(\frac{8}{5}-\frac{1}{2}\right)+\frac{41}{30}$$

$$=\frac{2}{3}\left(\frac{16-5}{10}\right)+\frac{41}{30}$$

$$=\frac{2}{3}\left(\frac{11}{10}\right)+\frac{41}{30}=\frac{11}{15}+\frac{41}{30}$$

$$=\frac{22+41}{30}=\frac{63}{30}=\frac{21}{10}$$

$$\frac{1}{2}\left(\frac{2y}{5}+3\right)=\frac{1}{2}\left(\frac{6}{5}+3\right)=\frac{1}{2}\left(\frac{21}{5}\right)=\frac{21}{10}$$

You have finished the program for Sec. 6-4. Now take the posttest to see how much you have learned.

POSTTEST 6-4

Find the simultaneous solution sets of the following equations. *Check your answers.*

1. $2x = 3y + 4$
   $12y = 7 + 8x$

2. $4x + 7y + 2 = 0$
   $-3y = 5x + 14$

3. $3x - 5y = 2$
   $10y = 6x - 4$

4. $\dfrac{7x}{5} + \dfrac{5y}{3} = -2$
   $\dfrac{9x}{4} - \dfrac{5y}{2} = -\dfrac{75}{4}$

5. $\dfrac{5x}{3} = \dfrac{3y}{7} + \dfrac{25}{21}$

$\dfrac{7x}{4} + \dfrac{9y}{8} - \dfrac{73}{8} = 0$

6. $\dfrac{3}{5}\left(\dfrac{3x}{2} - 3\right) + \dfrac{1}{2}\left(\dfrac{4y}{3} - 2\right) + \dfrac{151}{30} = 0$

$\dfrac{3}{4}\left(\dfrac{2x}{7} + 3\right) = \dfrac{5}{7}\left(\dfrac{y}{4} + 5\right) - \dfrac{33}{28}$

When you have finished the posttest, correct it yourself using the answers on page 332. Then check all the questions you got wrong on the list below:

1. ______ 3. ______ 5. ______

2. ______ 4. ______ 6. ______

If you compare your performance with that on Pretest 6-4, you will see how much you have improved. If you made more than one mistake on this posttest, review the material and take the pretest again.

Now proceed to Final Test 6.

FINAL TEST 6

Give the simultaneous solution sets of the following pairs of equations. Check your answers.

1. $x - 4y = 3$
   $x = -5$

2. $y = 5$
   $y = -3$

3. $x = -3$
   $2x = -6$

4. $y = 4x + 3$
   $5y - 6x = 9$

5. $y = 5 - 2x$
   $6x + 3y = 15$

6. $2x - 3y = -13$
   $10x + 7y = 12$

7. $7x + 5y = -16$
$3x - 8y = -17$

8. $3x - 7y = -5$
$6x + 2 = 14y$

9. $\frac{2x}{5} = \frac{5y}{3} - \frac{27}{20}$

$\frac{4x}{3} + \frac{7y}{4} - \frac{47}{48} = 0$

10. $\frac{3}{10}(x - 4) - \frac{2}{3}\left(\frac{y}{4} + 5\right) + \frac{89}{30} = 0$

$\frac{4}{5}\left(\frac{x}{3} + 2\right) = \frac{7}{20}(y + 4) + \frac{12}{5}$

The answers to this test are on page 332. Check the questions you got wrong on the list below and compare your results with those on Diagnostic Test 6 to see how much you have improved.

1. ______	4. ______	9. ______
2. ______	5. ______	10. ______
3. ______	6. ______	
	7. ______	
	8. ______	

If you made more than one mistake, review the material and then take the diagnostic test again. Make sure that you get all the questions right before you finish the chapter.

Now proceed to Chap. 7.

ANSWERS TO TESTS FOR CHAPTER 6

DIAGNOSTIC TEST 6

1. $\{(13,-3)\}$
2. $\emptyset$
3. $\{(x,y) \mid y = -2\}$
4. $\{(-1,-3)\}$
5. $\{(x,y) \mid y = 7 - 3x\}$
6. $\{(-\frac{5}{6}, \frac{4}{3})\}$
7. $\{(3,-1)\}$
8. $\emptyset$
9. $\{(\frac{1}{3}, -\frac{2}{3})\}$
10. $\{(-2,3)\}$

PRETEST 6-1

1.

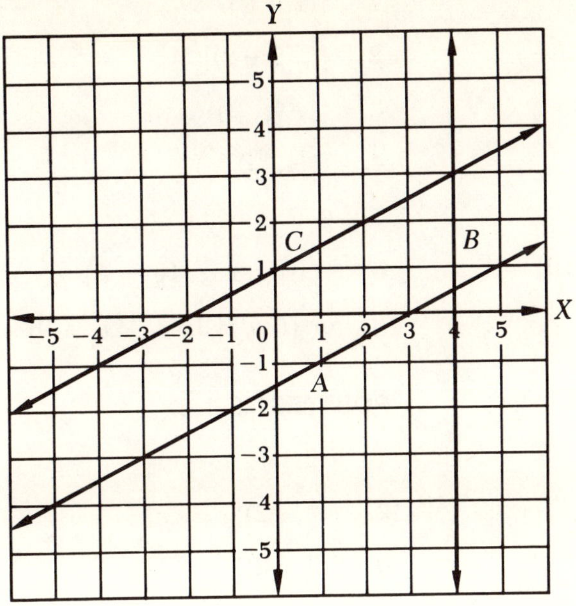

2. $\{(4, \frac{1}{2})\}$
3. $\emptyset$
4. $\{(4,3)\}$
5. $\{(2,-2)\}$
6. $\{(5, \frac{17}{5})\}$
7. $\emptyset$
8. $\{(x,y) \mid y = 2\}$

POSTTEST 6-1

1.

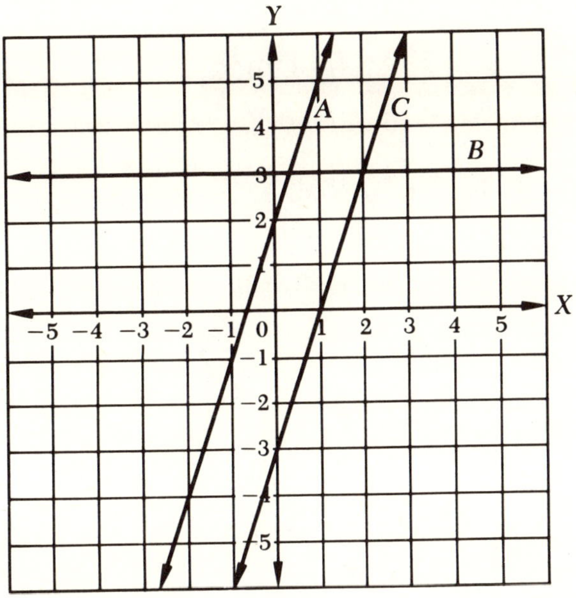

2. $\{(\frac{1}{3}, 3)\}$
3. $\emptyset$
4. $\{(2,3)\}$
5. $\{(-\frac{29}{2}, 5)\}$
6. $\{(-2, -\frac{5}{3})\}$
7. $\emptyset$
8. $\{(x,y) \mid y = 3\}$

PRETEST 6-2

1. $\{(2,3)\}$
2. $\{(1,3)\}$
3. $\{(\frac{1}{11},-\frac{4}{11})\}$
4. $\{(2,3)\}$
5. $\{(x,y) \mid 3y - 2x = 5\}$ or $\{(x,y) \mid 4x = 6y - 10\}$
6. $\emptyset$

POSTTEST 6-2

1. $\{(2,-1)\}$
2. $\{(-2,-2)\}$
3. $\{(-\frac{8}{13},\frac{59}{65})\}$
4. $\{(\frac{46}{17},\frac{60}{17})\}$
5. $\emptyset$
6. $\{(x,y) \mid x - 4y = 3\}$ or $\{(x,y) \mid 3x = 12y + 9\}$

PRETEST 6-3

1. $\{(1,-1)\}$
2. $\{(-2,3)\}$
3. $\{(1,2)\}$
4. $\{(\frac{4}{5},\frac{3}{5})\}$

POSTTEST 6-3

1. $\{(2,1)\}$
2. $\{(\frac{1}{2},-2)\}$
3. $\{(2,-3)\}$
4. $\{(\frac{3}{4},\frac{7}{4})\}$

PRETEST 6-4

1. $\{(x,y) \mid 2x - 3y = 5\}$
2. $\{(2,-3)\}$
3. $\emptyset$
4. $\{(3,4)\}$
5. $\{(-2,-1)\}$
6. $\{(-4,-6)\}$

POSTTEST 6-4

1. $\emptyset$
2. $\{(-4,2)\}$
3. $\{(x,y) \mid 3x - 5y = 2\}$
4. $\{(-5,3)\}$
5. $\{(2,5)\}$
6. $\{(-1,-2)\}$

FINAL TEST 6

1. $\{(-5,-2)\}$
2. $\emptyset$
3. $\{(x,y) \mid x = -3\}$
4. $\{(-\frac{3}{7},\frac{9}{7})\}$
5. $\{(x,y) \mid 2x + y = 5\}$
6. $\{(-\frac{5}{4},\frac{7}{2})\}$
7. $\{(-3,1)\}$
8. $\emptyset$
9. $\{(-\frac{1}{4},\frac{3}{4})\}$
10. $\{(3,-4)\}$

Chapter 7
Inequations of the First Degree in Two Variables

After Sec. 7-1 has showed you how the graphs of first-degree equations in two variables can be used as a basis for the graphs of the corresponding inequations, Sec. 7-2 introduces the graphing of the simultaneous solution sets of pairs of first-degree inequations in two variables.

This chapter requires knowledge of Chaps. 3, 5, and 6. Throughout this chapter the domain of each variable is assumed to be the set of real numbers.

DIAGNOSTIC TEST 7

Graph the solution sets of the following inequations:

1. $4x - 3y \leqslant -4$
2. $x + 2y > -2$

Graph the simultaneous solution sets of the following pairs of inequations:

3. $2x + y \leqslant 3$
   $2x - y \geqslant -3$
4. $3x + 2y \geqslant 4$
   $5x - 3y > -2$

When you have finished the diagnostic test, correct it yourself using the answers on page 361. Then check all the questions you got wrong on the list below:

1. ______
2. ______
3. ______
4. ______

If you made no mistakes at all, you need not do any work in this chapter and have finished this book. *Fundamentals of Intermediate Algebra* is available to improve your skills in more advanced algebra topics.

	section	page
If you made any mistakes in Question 1 or 2, circle Sec. 7-1.	7-1	335
If you made any mistakes in Question 3 or 4, circle Sec. 7-2.	7-2	348

You should now work through the sections whose numbers you have circled, taking them in numerical order. The sections begin on the pages indicated.

Section 7-1
Graphing Inequations

Do this section if you made a mistake in Question 1 or 2 of Diagnostic Test 7. Begin by taking the pretest below, which will show you how much of Sec. 7-1 you already know and what you will be able to do by the time you have finished it.

PRETEST 7-1

Graph the solution sets of the following inequations:

1. $2x + 5y > 4$
2. $x - 3y \leqslant 2$
3. $5x - 4y \geqslant -5$
4. $3x - 2y < -4$

When you have finished the pretest, correct it yourself using the answers on page 363. Then check all the questions you got wrong on the list below:

1. ______ 3. ______

2. ______ 4. ______

If you made any mistakes (even if it was only *one* mistake), read the introduction and go through the program.

If you made no mistakes at all, skip Sec. 7-1, because you already know its material, and go on to the next section whose number you circled on page 333. If no further numbers were circled, take Final Test 7 (page 360).

Introduction

In this section you will see that the solutions of an inequation in two variables, which is an open sentence, all lie on the same side of the line representing the corresponding equation. The graph of such an inequation is the area bounded by that line. This area may include or exclude points on the line itself. You will learn how to graph such areas by solving the corresponding equation, drawing the boundary line, and determining on which side of the line a given solution or nonsolution lies. The correct side of the line can easily be found by substituting (0,0) for (x,y), which shows whether or not the origin is a solution.

FRAMES FOR SECTION 7-1

1 Graph the solution sets of the following open sentences on sections of the number line:

a. $x + 3 > 6$
b. $x + 3 = 6$
c. $x + 3 \leq 6$

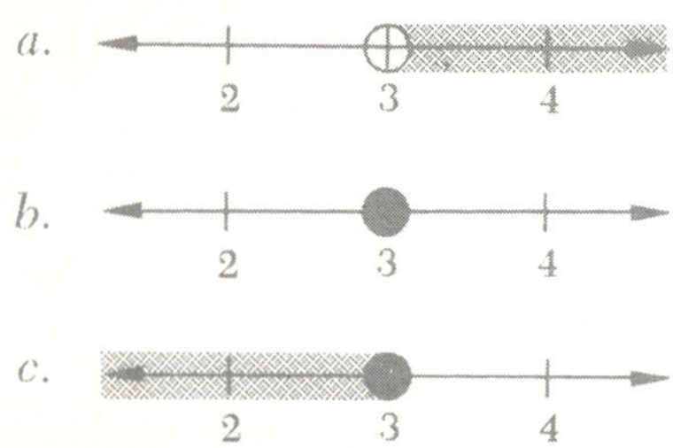

---

2 Graph the solution of $x + y = 5$ on a rectangular coordinate system.

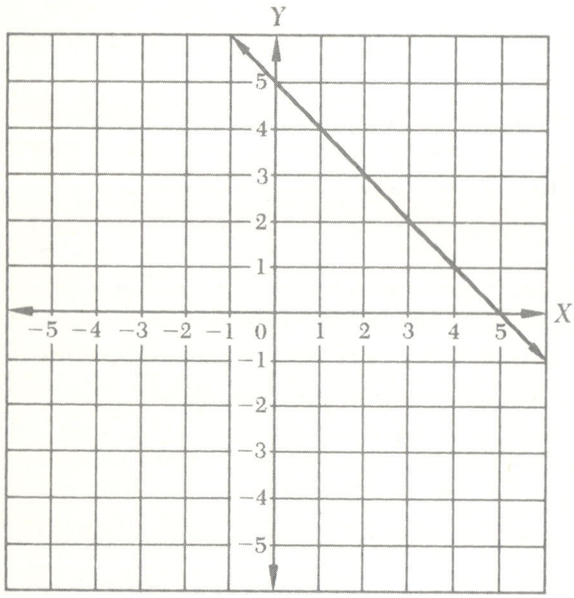

---

3 Sort the following ordered pairs into three categories:

a. Those which are solutions of $x + y < 5$.
b. Those which are solutions of $x + y = 5$.
c. Those which are solutions of $x + y > 5$.

Pair	A(0,5)	B(0,6)	C(−1,5)	D(2,4)	E(1,3)	F(2,3)	G(3,3)	H(3,1)	J(5,1)	K(6,−1)
Category	*b*									

A(0,5)	B(0,6)	C(−1,5)	D(2,4)	E(1,3)	F(2,3)	G(3,3)	H(3,1)	J(5,1)	K(6,−1)
b	*c*	*a*	*c*	*a*	*b*	*c*	*a*	*c*	*b*

---

4 *a.* Graph these 10 ordered pairs on the coordinate plane.
b. What property do pairs in category *b* have in common?
c. What property do pairs in category *a* have in common?
d. What property do pairs in category *c* have in common?

---

a.

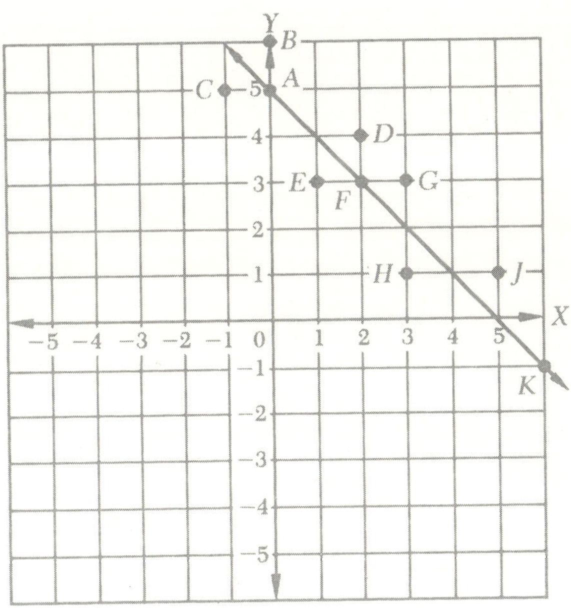

b. They are all on the graph of $x + y = 5$.
c. They are all on the bottom left side of the line $x + y = 5$.
d. They are all on the top right side of the line $x + y = 5$.

5 What area of the graph above represents the solution set of $x + y > 5$?

all the area on the top right side of the line

6 The graph of $x + y > 5$ is drawn as follows:

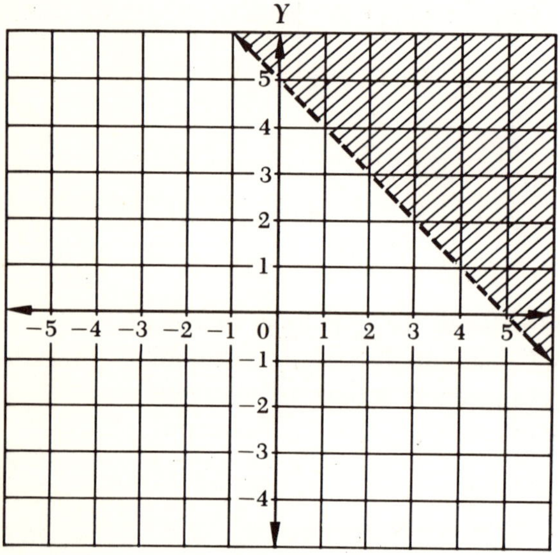

The dashed line indicates that the points on the line (are/are not) ______ solutions of $x + y > 5$. Draw the graph of $x + y < 5$.

are not

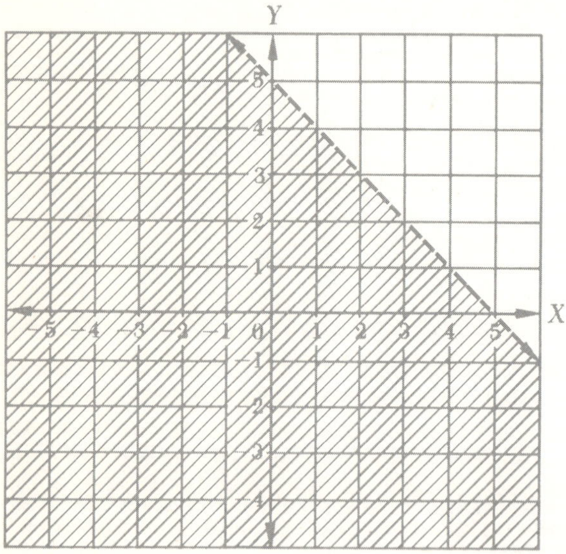

7 In the case of $x + y \leqslant 5$ the points on the line $x + y = 5$ (are/are not) ______ members of the solution set, and so a solid line is drawn to show that the points on the line (are/are not) ______ included. Which diagram represents the solution set of $x + y \leqslant 5$?

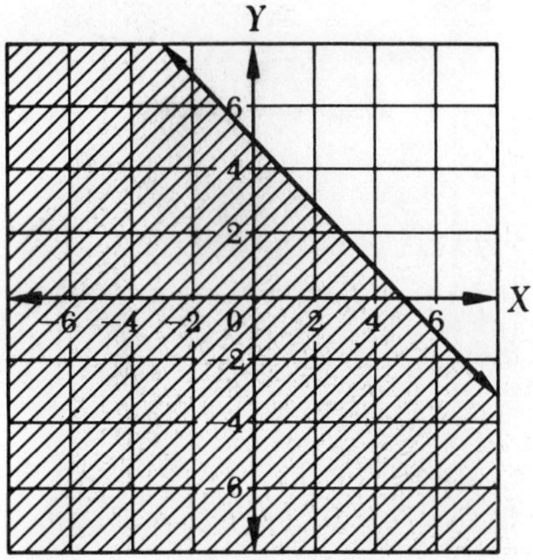

A

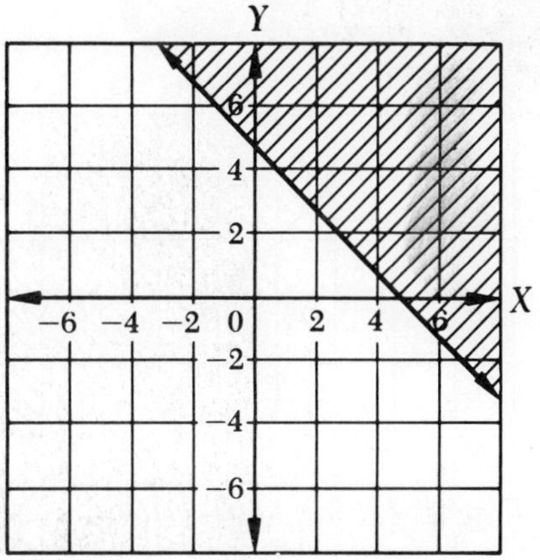

B

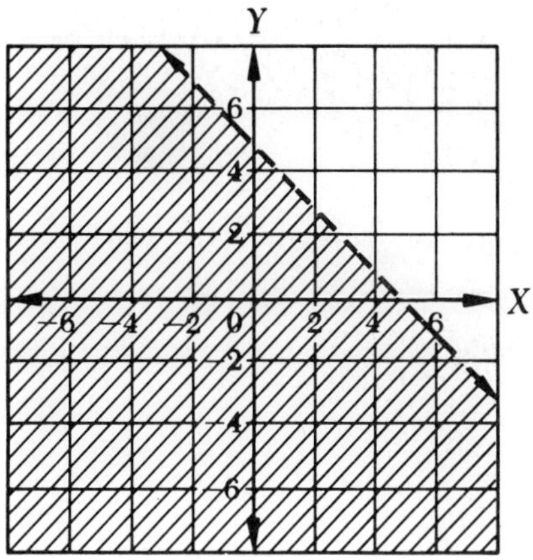

C

are; are; *A*

8 Which of the following open sentences is graphed in the diagram?

a. $x + y > 3$
b. $x + y < 3$
c. $x + y \geqslant 3$
d. $x + y \leqslant 3$

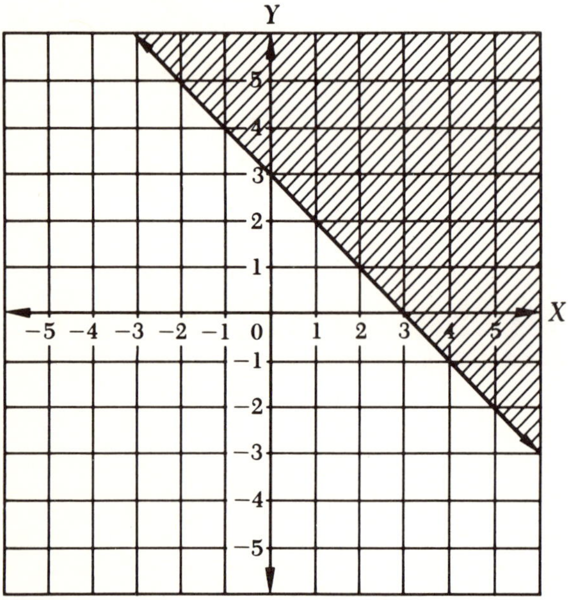

The solid line indicates that the points on the line (are/are not) ______ solutions.

---

c; are

---

9 Draw the graph of $x + y \leq 3$.

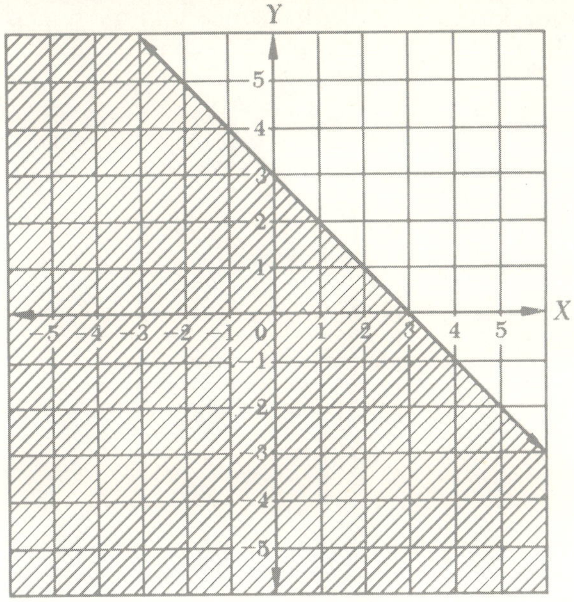

Make sure you have a solid line.

10 Draw the graph of $2y - x = 3$.

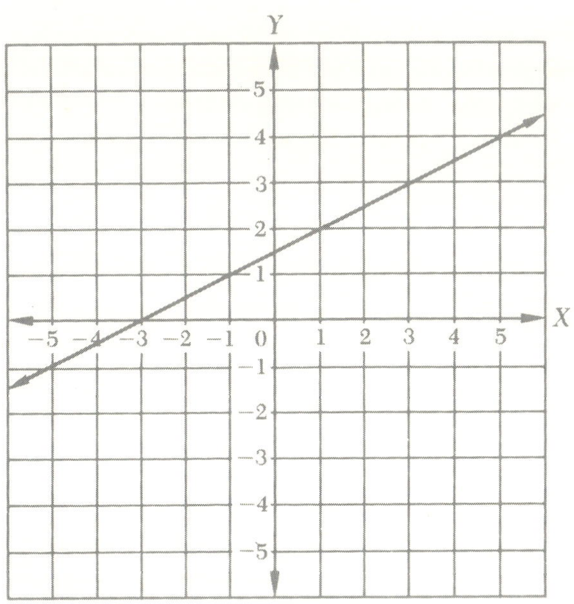

11 The origin (0,0) (is/is not) ______ a solution of $2y - x > 3$ because $2(0) - 0 > 3$ is a (true/false) ______ statement. Thus the solution set of $2y - x > 3$ is on the (same/opposite) ______ side of the line $2y - x = 3$ (as/to) ______ the origin. Graph the solution set of $2y - x > 3$.

---

is not; false; opposite; to

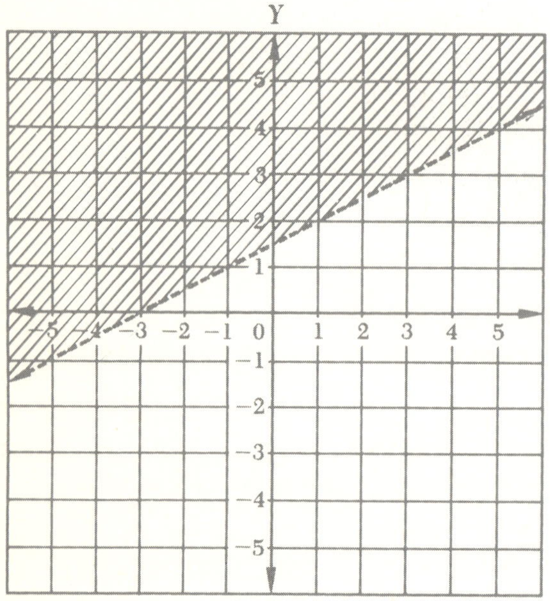

Make sure you drew a dotted line.

---

12 The origin (0,0) (is/is not) ______ a solution of $3x - y \leqslant -3$ because $3(0) - 0 \leqslant -3$ is a (true/false) ______ statement. Thus the solution set of $3x - y \leqslant -3$ (is/is not) ______ on the same side of the line $3x - y = -3$ as the origin. Graph the solution set of $3x - y \leqslant -3$.

---

is not; false; is not

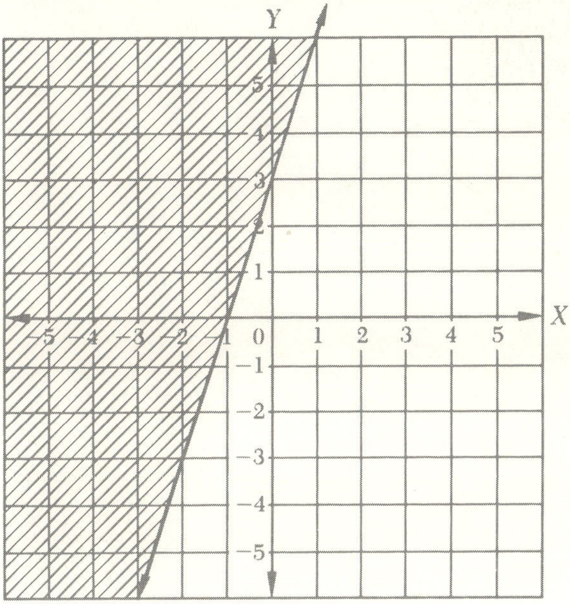

Make sure you drew a solid line.

---

13 Is the origin a solution of $3x + 2y > -2$? Graph the solution set of $3x + 2y > -2$.

---

yes

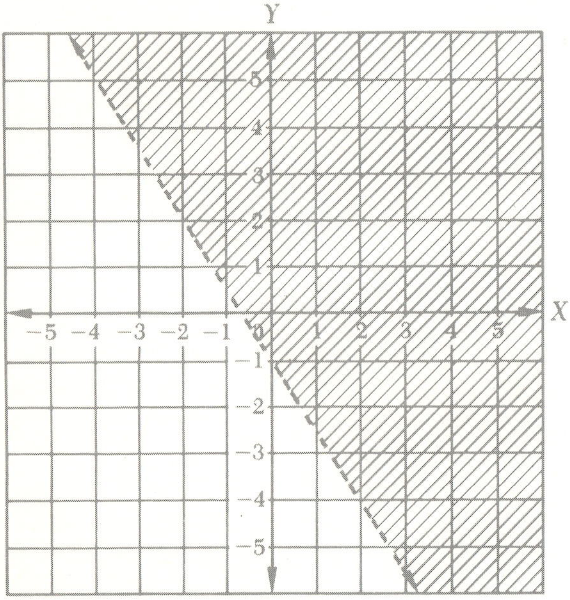

---

14 Is the origin a solution of $4x - 2y \geqslant -5$? Graph the solution set of $4x - 2y \geqslant -5$.

yes

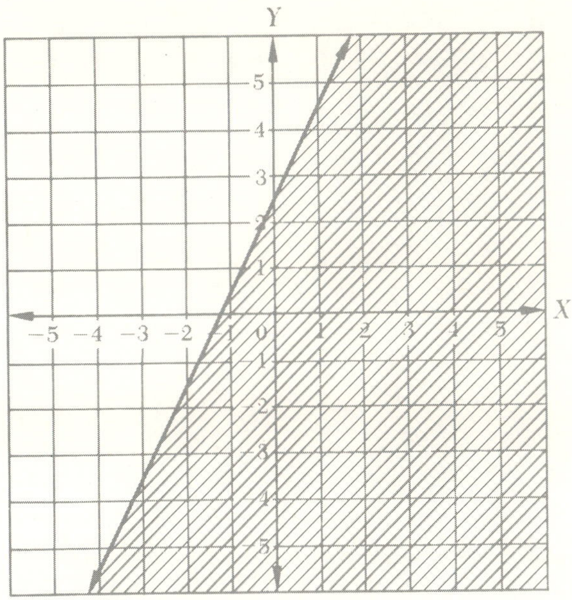

15 Graph the solution set of $3x + 4y < -3$.

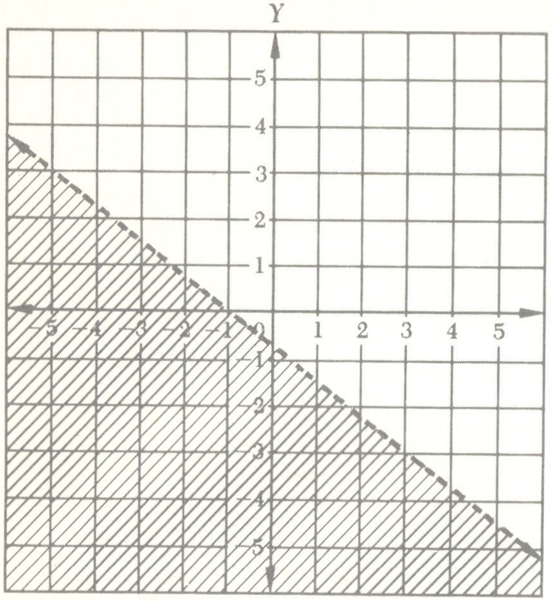

16 Graph the solution set of $2x - 5y \leqslant -1$.

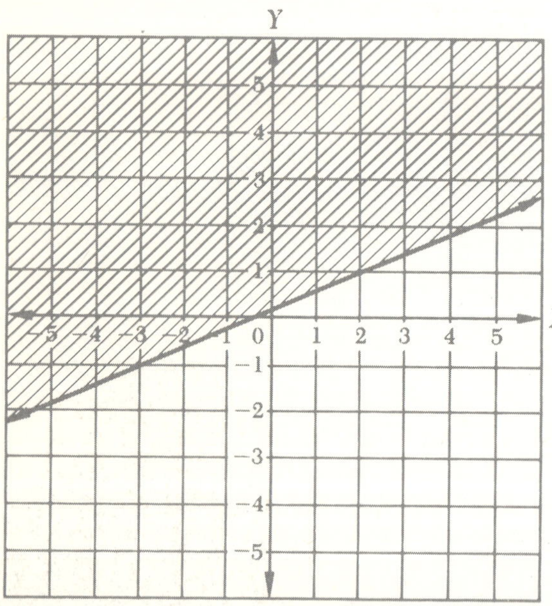

You have finished the program for Sec. 7-1. Now take the posttest to see how much you have learned.

POSTTEST 7-1

Graph the solution sets of the following inequations:

1. $3x + 5y < 6$
2. $x - 4y \geqslant 3$
3. $2x - 3y \leqslant -8$
4. $4x - 3y > -2$

When you have finished the posttest, correct it yourself using the answers on page 365. Then check all the questions you got wrong on the list below:

1. ______ 3. ______

2. ______ 4. ______

If you compare your performance with that on Pretest 7-1, you will see how much you have improved. If you made more than one mistake on this posttest, review the material and take the pretest again. Don't move on to the next section until you can get all the questions right.

Now proceed to the next section whose number you circled on page 333. If no other sections were circled, take Final Test 7 (page 360).

Section 7-2
Graphing Simultaneous Solution Sets of Pairs of Inequations

Do this section if you made a mistake in Question 3 or 4 of Diagnostic Test 7. Begin by taking the pretest below, which will show you how much of Sec. 7-2 you already know and what you will be able to do by the time you have finished it.

PRETEST 7-2

Graph the simultaneous solution sets of the following inequations:

1. $4y - 3x \leqslant 2$
   $2y + 5x > -7$

2. $2x + 4y \geqslant 3$
   $5x - 3y > -4$

3. $2x - 7y < -4$
   $4x - 2y \geqslant 5$

When you have finished the pretest, correct it yourself using the answers on page 367. Then check all the questions you got wrong on the list below:

1. ______
2. ______
3. ______

If you made any mistakes (even if it was only *one* mistake), read the introduction and go through the program.

If you made no mistakes at all, skip Sec. 7-2, because you already know its material, and go on to take Final Test 7 (page 360).

Introduction

After you are shown that the simultaneous solution set of a pair of inequations in two variables is the intersection of the solution sets of the equations, you see that graphically this is represented by the area common to both solution sets. You check the simultaneous solution set by testing a point in the proposed solution area.

FRAMES FOR SECTION 7-2

1 Graph the equations $x + y = 2$ and $x - y = 3$ on the same coordinate system.

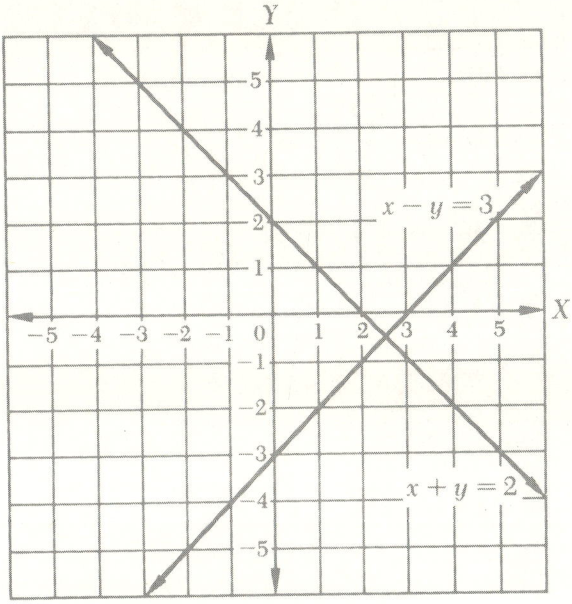

2 Insert < or > to make the statements correct:

a. The point (4,0) is a solution of $x + y$ ____ 2 and of $x - y$ ____ 3.

b. The point (0,−4) is a solution of $x + y$ ____ 2 and of $x - y$ ____ 3.

a. >; > *b.* <; >

3 The diagram below shows the simultaneous solution set of $x + y$ ______ 2 and $x - y$ ______ 3 (insert < or >).

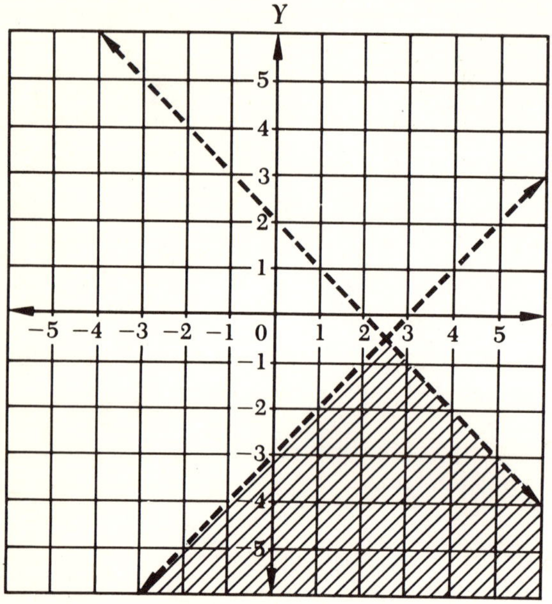

< >

---

4 Graph the solution set of $2x + y \geqslant -3$ and the solution set of $y - 2x < 4$ on separate coordinate systems.

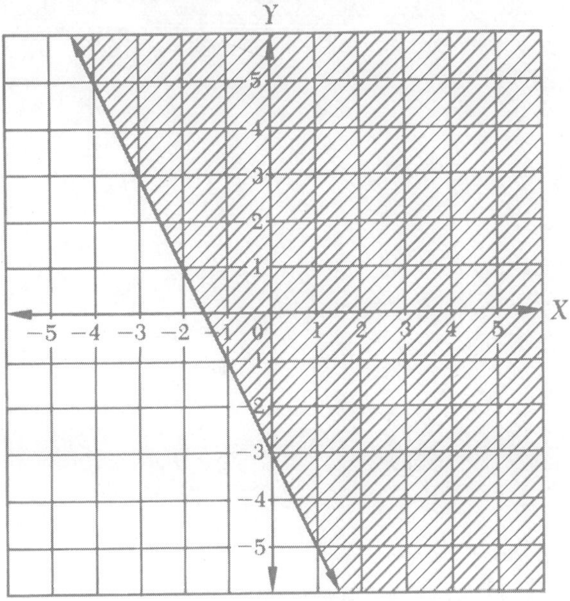

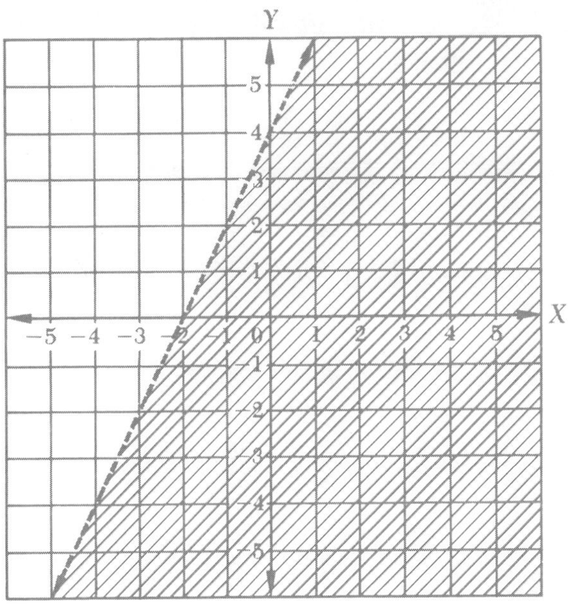

5 Graph the simultaneous solution set of $2x + y \geqslant -3$ and $y - 2x < 4$. Check your answer by testing a point in the area concerned.

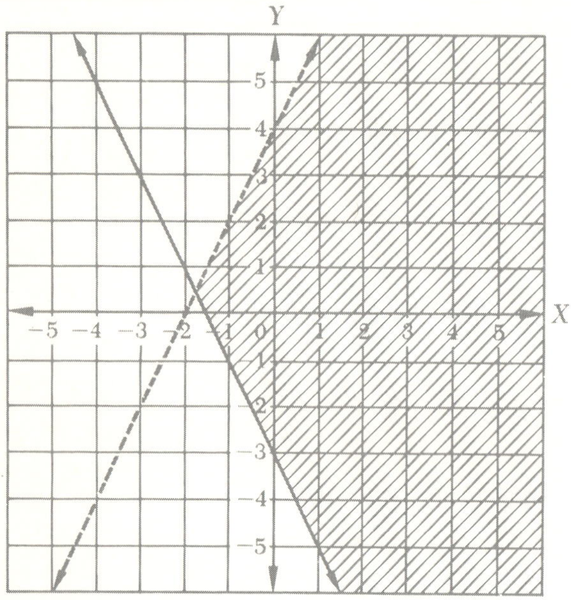

CHECK

For point (0,0)

$0 + 0 \geqslant -3$
$0 - 0 < 4$

You could have chosen any other point in the solution area to test your answer.

---

6 Graph the solution set of $2x + 5y > 2$ and the solution set of $y - 4x \geqslant 3$ on separate coordinate systems.

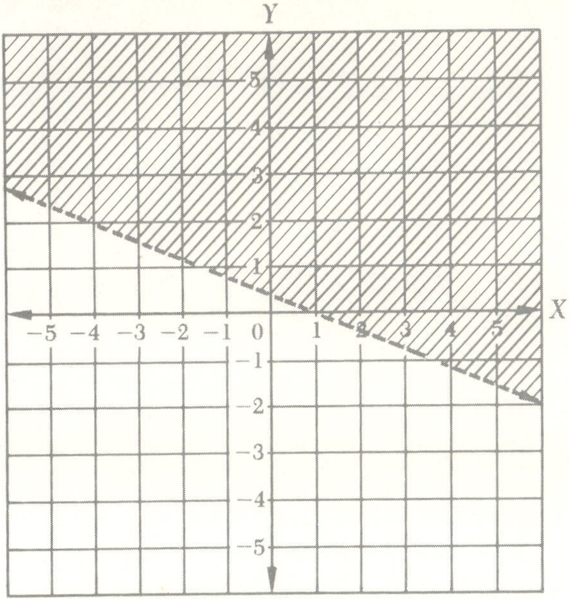

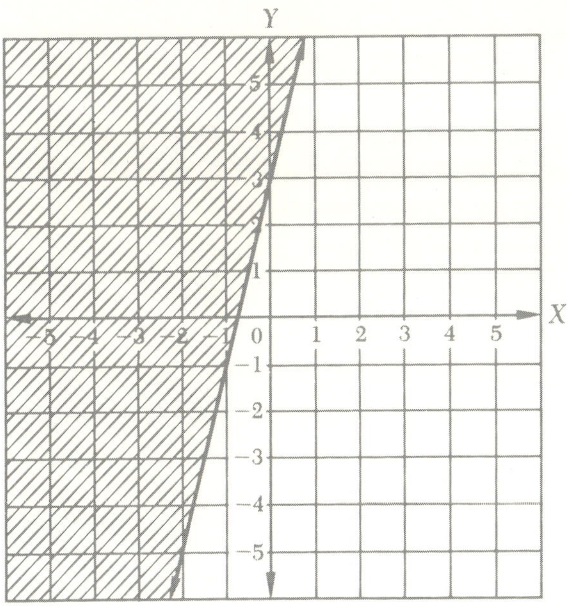

7 Graph the simultaneous solution set of $2x + 5y > 2$ and $y - 4x \geqslant 3$. Check your answer by testing a point in the area concerned.

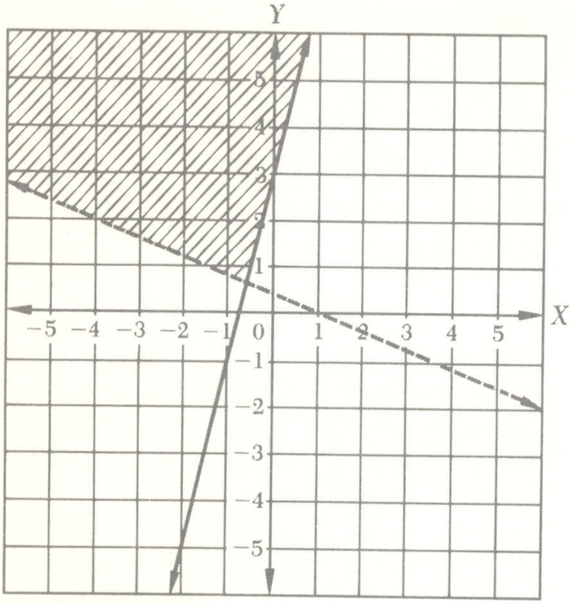

CHECK

For point $(-1, 1)$

$$2(-1) + 5(1) = -2 + 5 = 3 > 2$$
$$1 - 4(-1) = 1 + 4 = 5 \geqslant 3$$

8 The diagram below shows the graphs of $2x + 3y = 5$ and $3y - 2x = -3$.

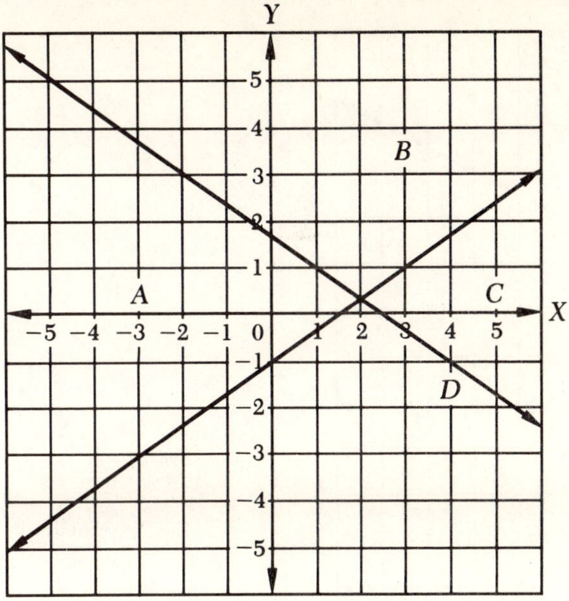

Insert < or > to make the following statements correct:

a. (0,0) is a solution of $2x + 3y$ _____ 5 and of $3y - 2x$ _____ −3.

b. Therefore area *A* is the graph of the simultaneous solution set of $2x + 3y$ _____ 5 and $3y - 2x$ _____ −3.

c. Area *C* (opposite area *A*) is the graph of the simultaneous solution set of $2x + 3y$ _____ 5 and $3y - 2x$ _____ −3.

---

a. <; > *b.* <; > *c.* >; <

---

9 Now let us choose a second point to distinguish between *B* and *D*:

a. (0, 4) is a solution of $2x + 3y$ _____ 5, and of $3y - 2x$ _____ −3.

b. Area *B* is the graph of the simultaneous solution set of ________ and ________.

c. Area *D* is the graph of the simultaneous solution set of ________ and ________.

---

a. >; > *b.* $2x + 3y > 5$; $3y - 2x > -3$
c. $2x + 3y < 5$; $3y - 2x < -3$

---

10 The diagram below shows the graphs of $4x - 2y = -5$ and $2y - x = 3$.

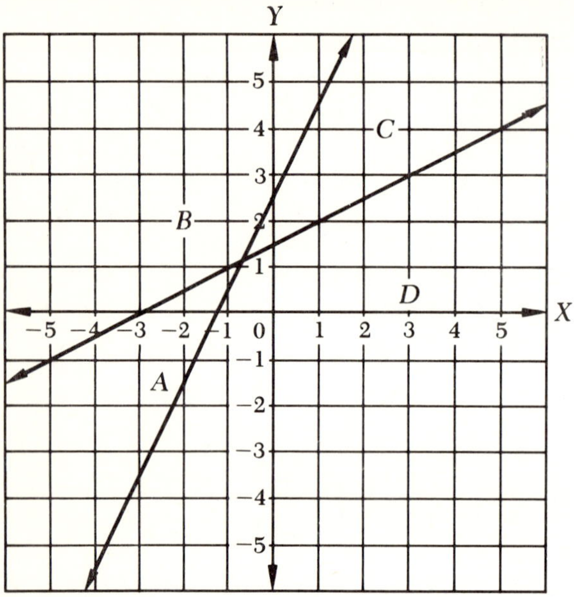

a. Area D is the simultaneous solution set of $4x - 2y$ ______ -5 and $2y - x$ ______ 3.

b. Area B is the simultaneous solution set of __________ and __________.

c. Area A is the simultaneous solution set of __________ and __________.

d. Area C is the simultaneous solution set of __________ and __________.

---

a. $>$; $<$

b. $4x - 2y < -5$; $2y - x > 3$

c. $4x - 2y < -5$; $2y - x < 3$

d. $4x - 2y > -5$; $2y - x > 3$

---

11 Graph the simultaneous solution set of $x + 2y > 3$ and $y - 3x \leq -5$. Check your answer by testing a point in the area concerned.

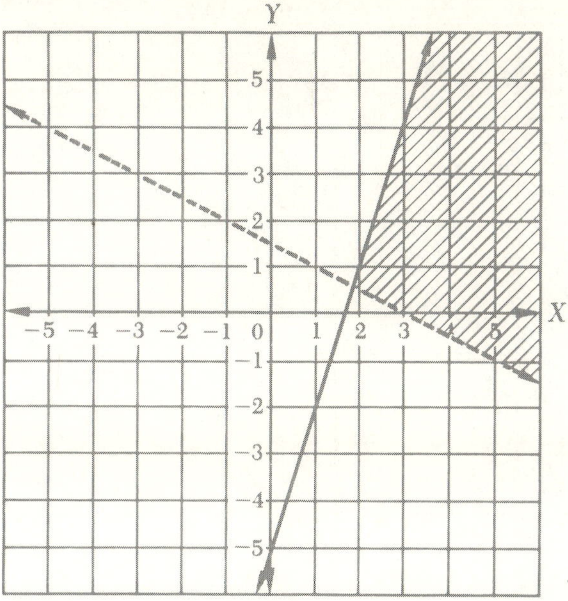

CHECK

For point (4,0)

$4 + 0 > 3$
$0 - 12 \leq -5$

12 Graph the simultaneous solution set of $3y + x \leqslant 4$ and $2x - 3y < 4$. Check your answer by testing a point in the area concerned.

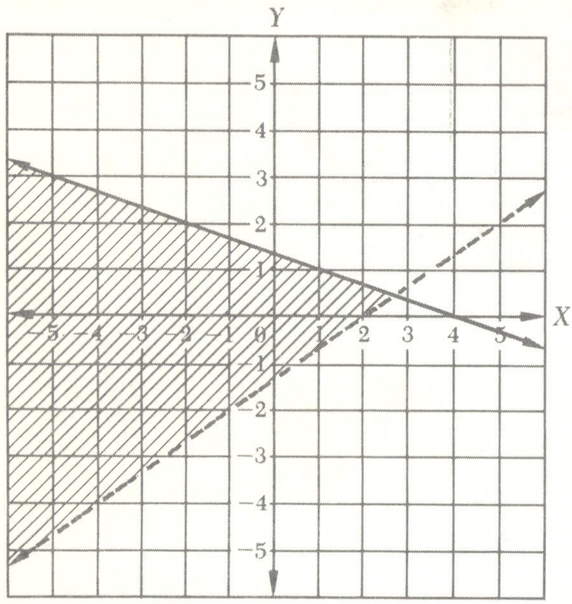

CHECK

For point (0,0)

$0 + 0 \leqslant 4$
$0 - 0 < 4$

You have finished the program for Sec. 7-2. Now take the posttest to see how much you have learned.

POSTTEST 7-2

Graph the simultaneous solution sets of the following pairs of inequations:

1. $5y - 4x \leqslant 3$
   $3y + 2x > -7$

2. $3x + 5y \leqslant 2$
   $4x - 3y < -5$

3. $6x - y > 5$
   $2x - 7y \geqslant 4$

When you have finished the posttest, correct it yourself using the answers on page 368. Then check all the questions you got wrong on the list below:

1. ______
2. ______
3. ______

If you compare your performance with that on Pretest 7-2, you will see how much you have improved. If you made more than one mistake on this posttest, review the material and take the pretest again.

Now proceed to Final Test 7.

FINAL TEST 7

Graph the solution sets of the following inequations:

1. $3x - 4y \leqslant -4$

2. $2x + 3y > -2$

Graph the simultaneous solution sets of the following pairs of inequations:

3. $2x + y > 3$
   $3x - 2y \leqslant -2$

4. $3x + 4y \geqslant 6$
   $2x - 3y < -4$

The answers to this test are on page 370. Check the questions you got wrong on the list below and compare your results with those on Diagnostic Test 7 to see how much you have improved.

1. ______ 3. ______

2. ______ 4. ______

If you made more than one mistake, review the material and then take the diagnostic test again. Make sure that you can get all the questions right before you finish the chapter.

Now that you have finished this book, you may want to improve your skills in more advanced algebra topics. *Fundamentals of Intermediate Algebra* is available for this purpose.

ANSWERS TO TESTS FOR CHAPTER 7

It is suggested that you use the marked points to check the accuracy of your graph.

DIAGNOSTIC TEST 7

1.

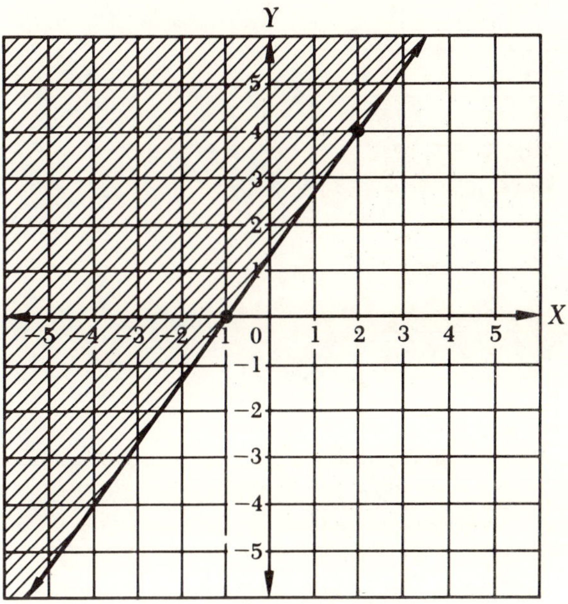

2.

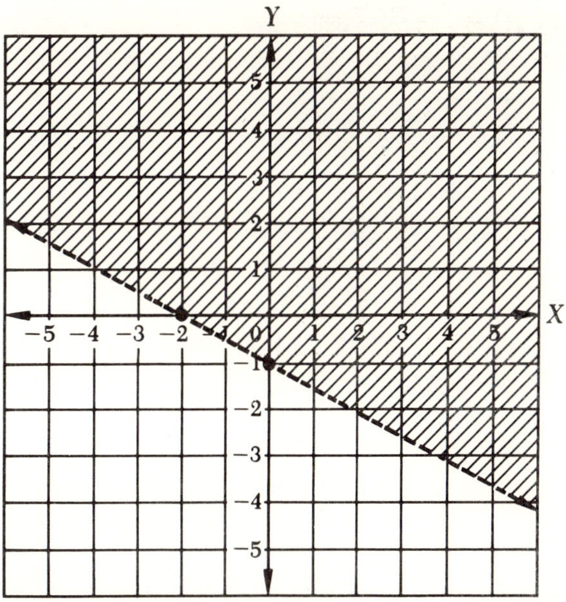

3.

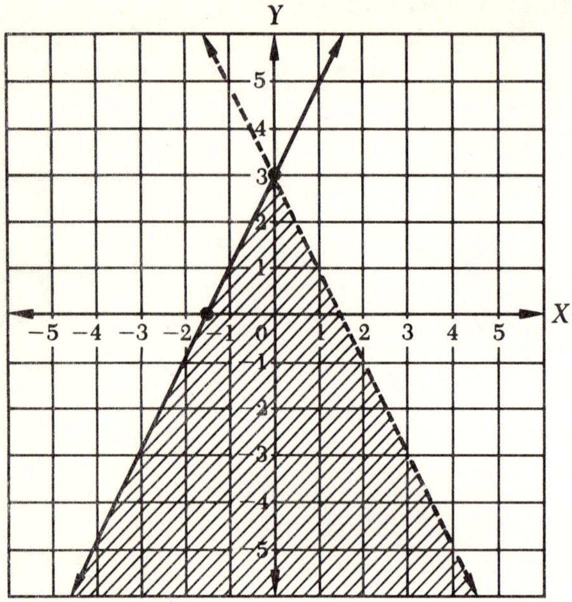

4.

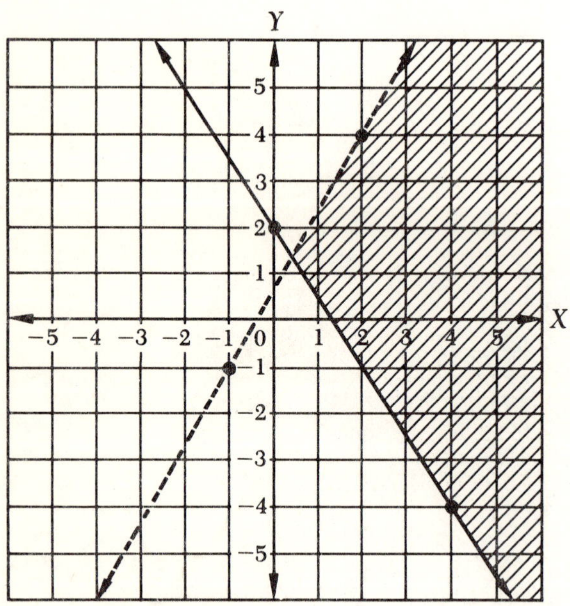

PRETEST 7-1

1.

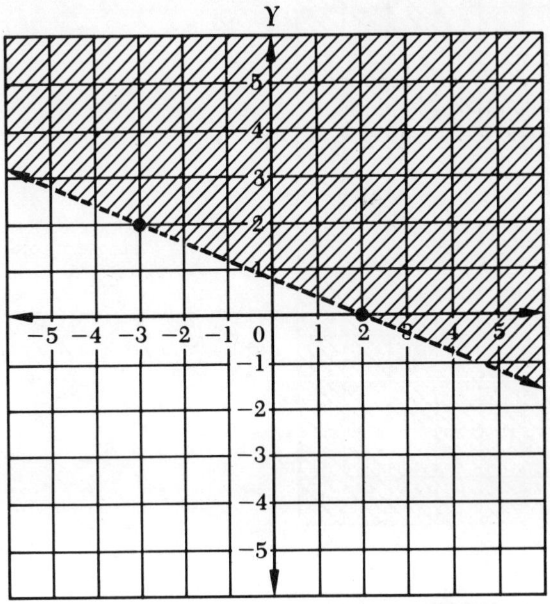

2.

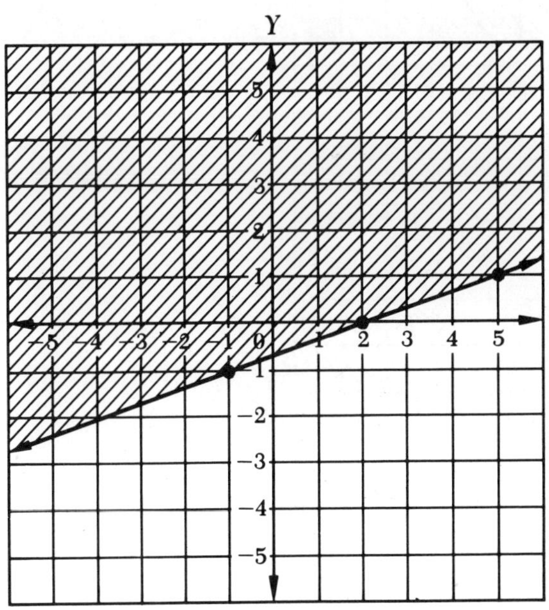

3.

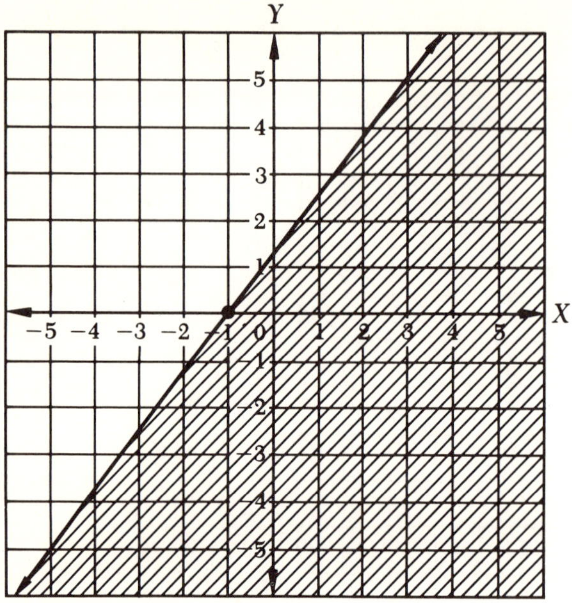

4.

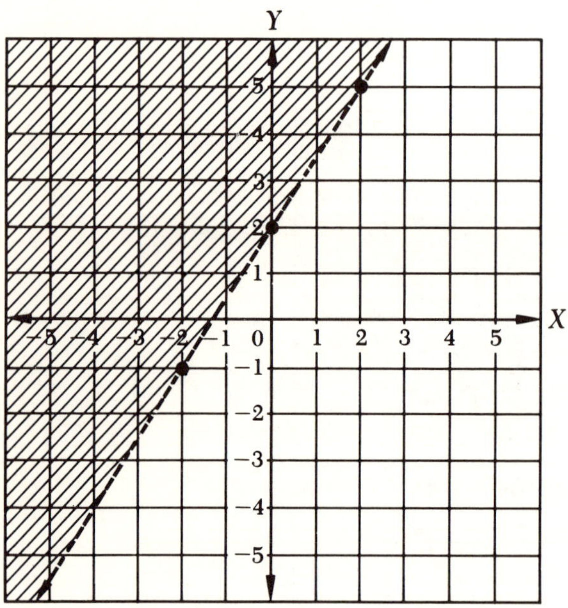

POSTTEST 7-1

1.

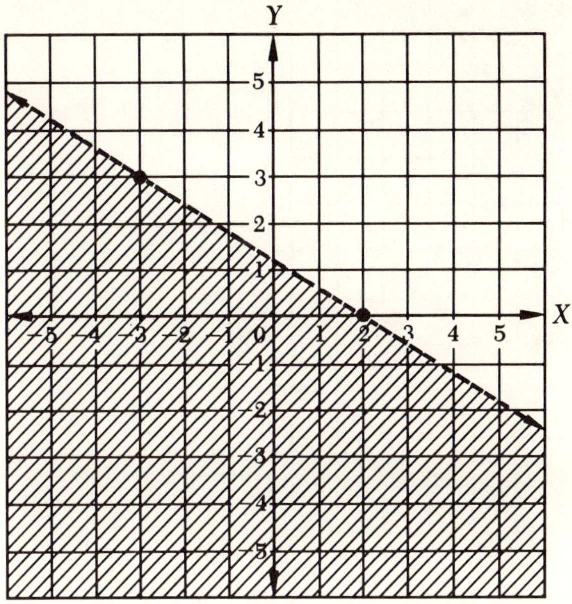

2.

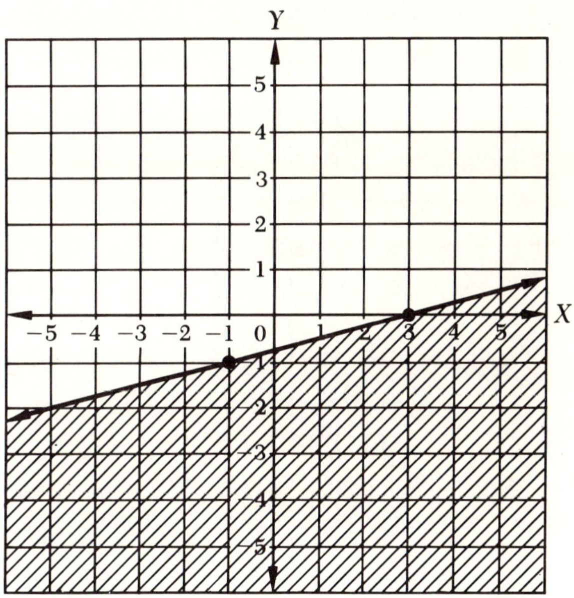

3.

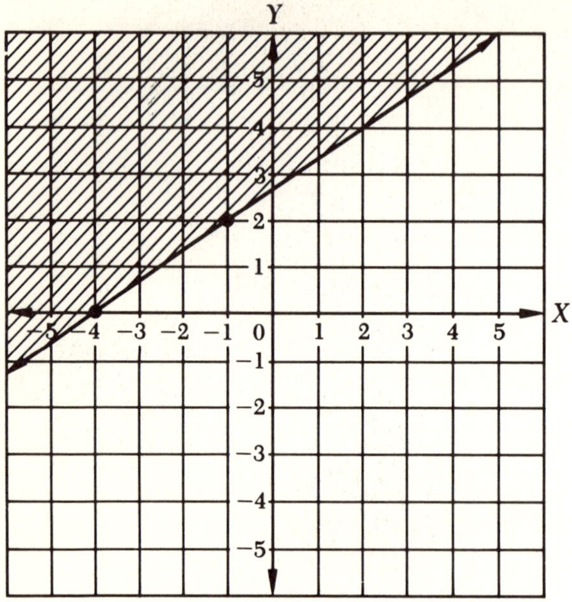

4.

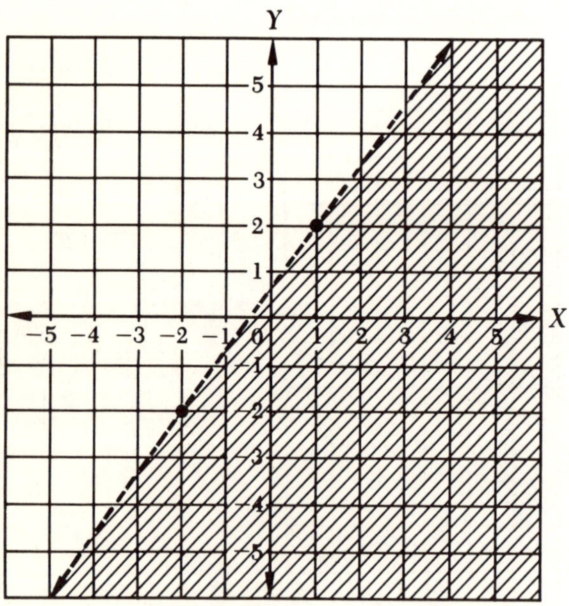

PRETEST 7-2

1.

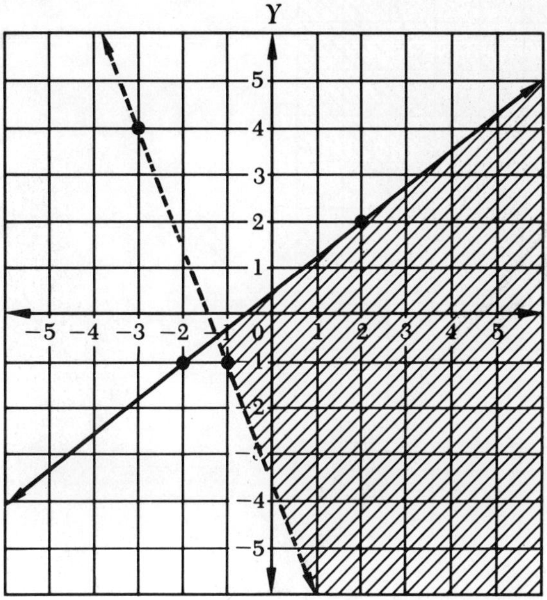

2.

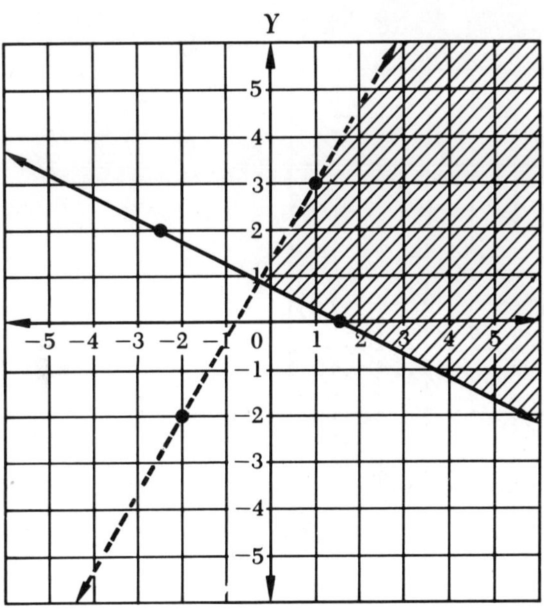

3.

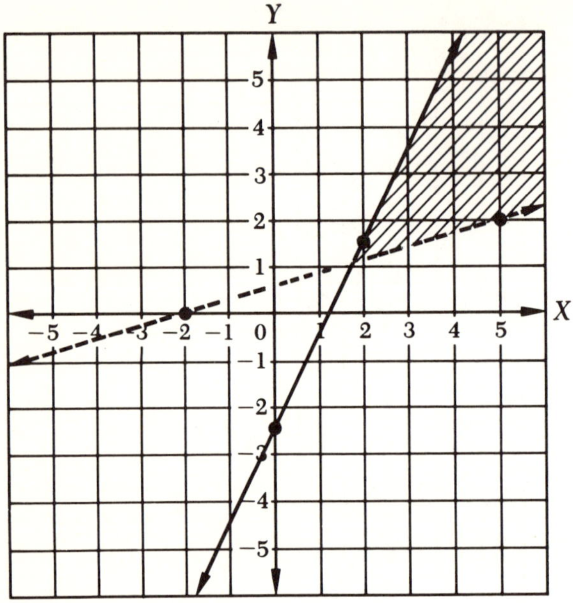

POSTTEST 7-2

1.

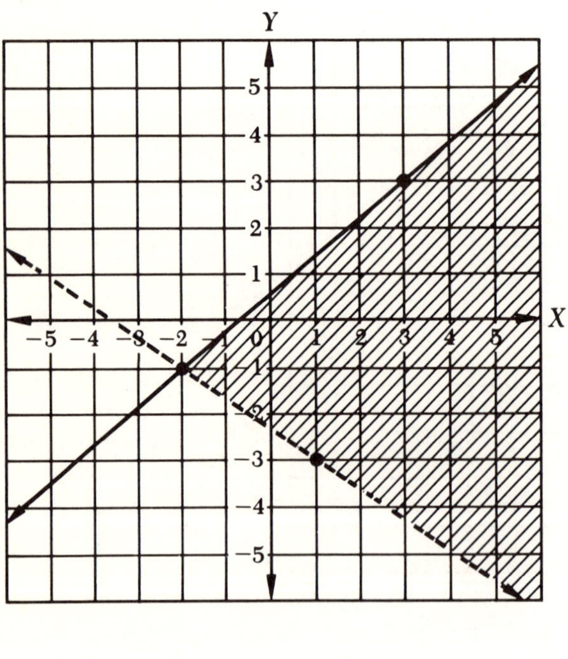

2.

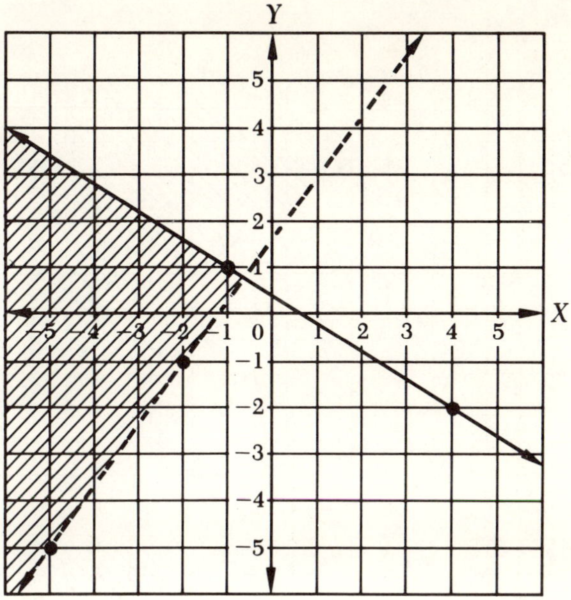

3.

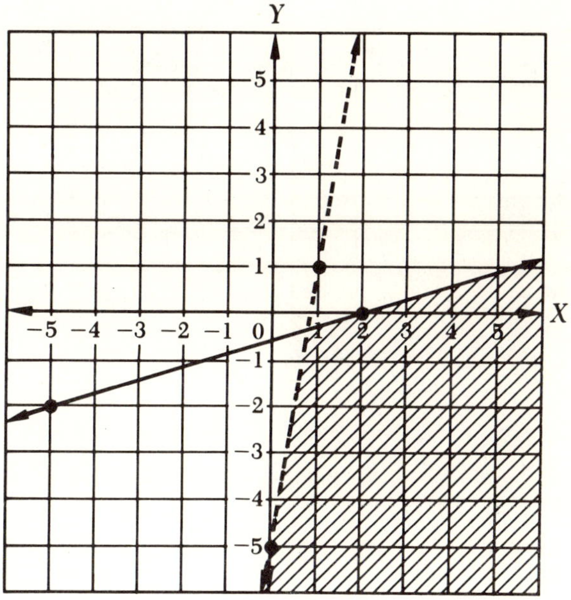

FINAL TEST 7

1.

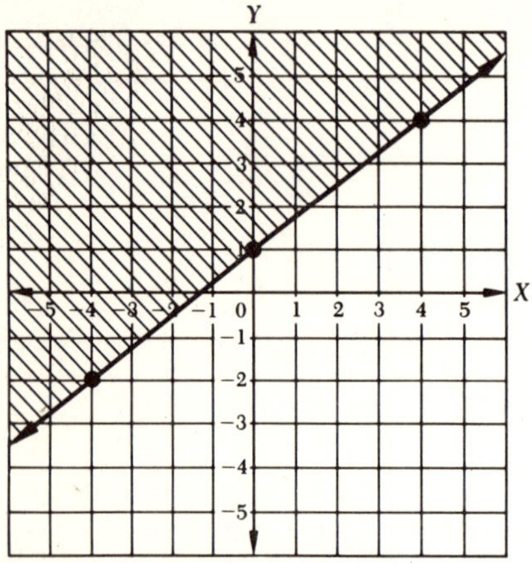

2.

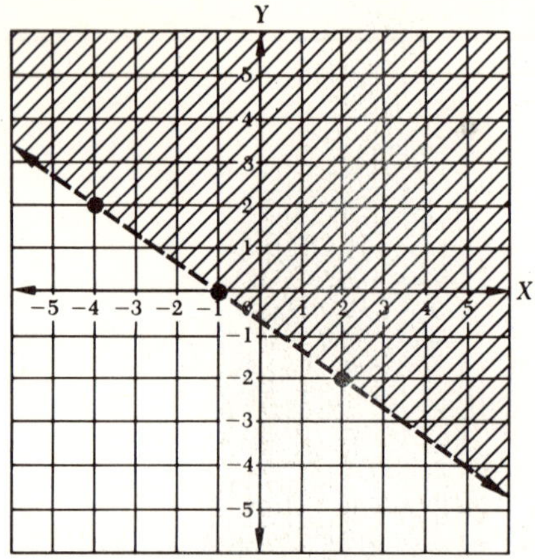

3.

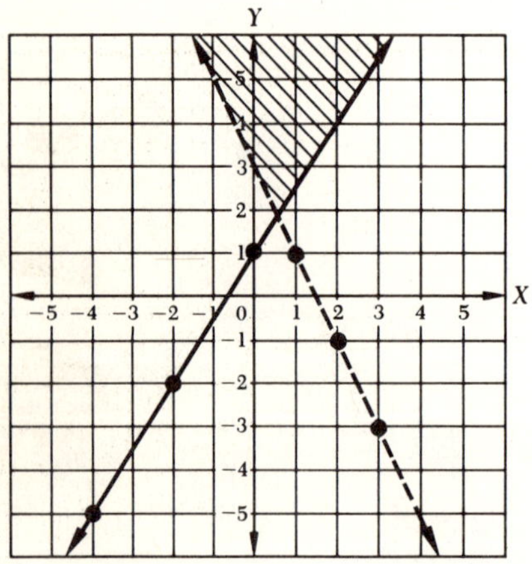

4.

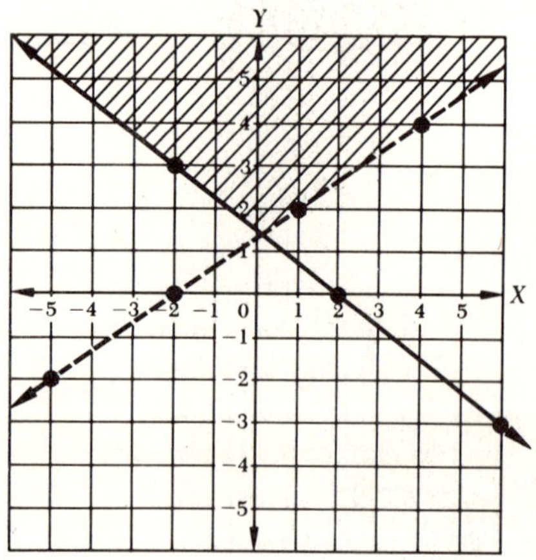